Acoustics in Performance and Worship

Richard A. Honeycutt, Ph.D.

Parson's Porch

Parson's Porch Books

Acoustics in Performance and Worship

This book was printed in the United States of America.

To order additional copies of this book, contact:

Parson's Porch & Company

1-423-475-7308

www.parsonsporchbooks.com

Table of Contents

III. Attaining the Goal

V. Controlled Acoustics

VI. Working with Your Room

VII. References

Introduction

For far too long, acoustical design has been popularly considered at worst a gamble, and at best a black art. If this position has ever been defensible, it certainly is not so today. In his 1932 book on the subject, Dr, Vern O. Knudsen, Professor of Physics and subsequently chancellor of UCLA, wrote:

> The idea, still shared by some architects, builders, and ... authorities, that the acoustical outcome of a ... building cannot be determined until the building is completed is an untenable one and can no longer be used as an excuse for poor acoustics. The acoustical outcome of a ... building is a problem in good designing and good engineering, and if the fundamental principles of architectural acoustics are incorporated in the design of the building there need be no uncertainty as to the acoustical outcome of that building – the acoustics will be good. If these principles are not incorporated, or if they are violated, there likewise need be no uncertainty – the acoustics will be bad, bad to the degree that the principles have been ignored or violated. (Knudsen, 1932)

Yet eighty-one years later, we still find that buildings with good acoustics are the exception rather than the rule. In seeking a reason why this is so, one is tempted to point the finger at architects, who have little or no training in acoustics, and may overestimate the extent of acoustical knowledge they do possess. However, as in most economic markets, the clients of architects get what they demand. If a client insists on good acoustics from the inception of a project, most architects will engage the services of a competent acoustical consultant. And if the client remembers that most things worth having come with a cost, design features recommended by the acoustical consultant will be more likely to survive cost-cutting and other compromises of the building process. Finally, if communication of the consultant's recommendations to the building contractor is given a high priority, those recommendations stand a good chance of being incorporated into the building – and good acoustics will be the result.

The purpose of this book is to present those responsible for providing good acoustics in performance and worship spaces an understanding of the variables and choices entailed in proper acoustic design for worship. The readership is expected to include architects, pastors, ministers of music, technical teams, and musicians. Practicing acoustical consultants may find the book a useful reference as well. The author hopes that the readers will find the level of presentation comfortable and straightforward without being simplistic.

--Richard A. Honeycutt, Ph.D.

Acknowledgments

I would like to thank many people for their help in providing background information that I needed in writing this book. I have learned much from Dr. Noral Stewart in the four+ decades of our acquaintance, and he has also served as a reviewer of this volume. Dr. Linda Franzoni and Dr. Don Bliss of Duke University have been great teachers and exemplars of acoustical research and practice. Dr. Michael Patton of The Union Institute and University instilled in me a sense of the importance of careful cross-checking of secondary references before quoting an "authority". The late Dr. Richard Campbell, adjunct faculty of M. I. T. and W. P. I. has provided valuable guidance on the practice of architectural acoustics. Dr. Bengt-Inge Dalenback has been a long-distance friend and a wonderful resource not only on computerized acoustical modeling of architectural spaces, but also on architectural acoustics in general. Joe Bridger has shared nuggets from his many years of experience in architectural acoustics and noise control. Dr. John Ferguson of the music faculty of St. Olaf's College has been most encouraging. The late Paul W. Kirsch has inspired me greatly in the science of electroacoustics, both through his research and writing, and through our occasional personal meetings. Dr. Jo Ann Poston, founding director of the Lexington Choral Society, and a career music educator and church musician, has been of great service as a reviewer. And my wife Betty Jane has provided unflagging emotional support through the long years of manuscript preparation.

To all these many benefactors I offer my most sincere gratitude.

Richard A. Honeycutt, Ph.D.

I. The Science of Acoustics

1. Architectural Acoustics – Then and Now

Acoustics is a very ancient science. The amphitheaters of southern Europe furnish the subject matter for our study of earliest acoustical structures. At first, a small group of listeners could congregate around a speaker on a level plane. With the development of public oratory and drama, more seats were added, and the speaker was raised above the listeners to provide them a better view. A commonly available structure for this purpose was the threshing floor. Later, the more distant seats were raised to better enable their listeners to hear. This is especially important in a warm climate where temperature gradients can cause the sound to curve upward as it progresses. (This effect is only significant at large distances from the source (>100m). Seats were arranged in a semicircular pattern for greatest visual and aural intimacy. The absence of walls and roof was a mixed blessing: an amphitheater is not subject to echoes or to excessive reverberation, but neither can it offer the benefits of helpful reflections that reinforce the voice of the speaker or the sound and timbre of musical instruments. Consequently, Greek dramatists used elaborate masks with the grossly enlarged mouths acting as megaphones to better project the voice. Greek architects developed the use of clay and bronze resonant pots located at specific positions to reinforce the sound. This practice is described by Marcus Vitruvius Pollio, first-century Roman architect, in Book 5 of his "Ten Books on Architecture." Vitruvius's book discusses such acoustical phenomena as "dissonance", which we would call acoustical interference; "circumsonance" (reverberation); and "resonance" (echoes), in addition to the "sounding vessels."

In more northerly regions, audience comfort demanded the use of enclosed spaces, and Turkish buildings of the Ottoman and earlier periods employed resonant cavities to help control excessive reverberation and standing waves, making speech more intelligible. The great cathedrals of medieval Europe were not designed for speech intelligibility, but according to aesthetic and symbolic concerns. The result was buildings that provided maximum reinforcement of chant, coupled with extreme reverberation that allows the development of harmony resulting from the interplay of the time sequence of tones in a monophonic chant. Speech intelligibility in the cathedrals was virtually nil.

During the nineteenth century, efforts toward acoustic enhancement of auditoria and concert halls were principally directed toward providing reflecting surfaces ("sounding boards") to concentrate the energy from a speaker, singer, or instrument(s) toward the audience. In 1853, Dr, J. B. Upham of Boston undertook a careful study of the acoustics of the Boston Music Hall, culminating in an essay discussing reverberation and resonance.

American physicist Joseph Henry presented papers to the American Association for the Advancement of Science in 1854 and 1856, and a new lecture room in the Smithsonian Institution was built according to the results of his experiments and theory. The results were said to be highly satisfactory.

Architect T. Roger Smith published his book *Acoustics of Public Buildings* in 1861, in which he discussed the differing acoustical requirements of speech and music. Tyndall and Rayleigh investigated the control of reverberation in rooms. But the definitive work that is universally agreed to mark the genesis of scientific acoustics was that of Harvard University Physics Professor Wallace Clement Sabine. In the very last years of the nineteenth century, Sabine was tasked with remedying the horrific acoustics of the Fogg Lecture Hall. He began with a quantitative study of the nature of reverberation and the physical factors controlling it. His equation for predicting the reverberation time of a room is still widely used. But perhaps even more important was the widespread realization, upon publication of his work, that acoustics need no longer be considered a black art, but could and indeed should be studied among the physical and engineering sciences. The scientific history of the twentieth century is replete with advances made by Sabine's successors, who have advanced analytical and predictive architectural acoustics to its current state of accuracy and utility.

2. Physical Acoustics: The Nature of Sound

Let us suppose that an object is vibrating; that is, moving back and forth in a regular manner. If we graph the displacement of the object from its central position *vs.* time, the graph might look something like the one shown in Figure 2-1. In this figure, the displacement axis is vertical, and the time axis is horizontal. Zero displacement, or a point at the vertical center of the graph, represents the object when it is located at the central position of its vibration. Zero time is when the vibration begins, and as time increases, we move toward the right-hand end of the graph.

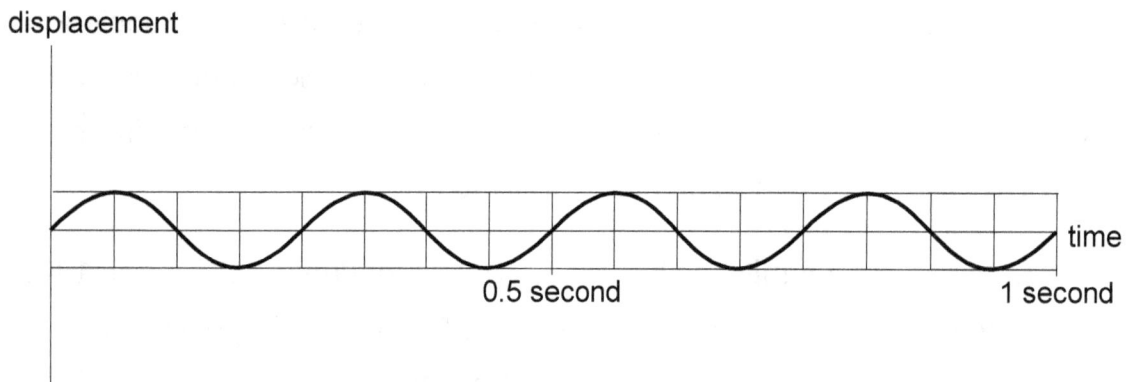

Figure 2-1: Vibratory Motion

This simplest form of vibratory motion is what occurs when a mass is suspended from a spring, then deflected downward and released. Our graph shows a regular alternation, or **oscillation**, between a certain positive (upward) maximum displacement and that same amount of negative (downward) displacement, which is what would occur if there is no friction – either from air surrounding the spring and mass or internal friction in the spring. The shape of this graph is exactly the same at that of the sine function, plotted on the vertical axis against the value of angle on the horizontal axis, and so it is often called a **sine wave**.

Notice that our sine wave graph shows four complete zero-to-positive-maximum-to-zero-to negative-maximum-to zero cycles. These four cycles occur in a one-second interval of time. Thus we can say that the **frequency** of the oscillation is four cycles per second, or four **hertz (Hz)**.

Another way to look at frequency is to say that if there are four cycles per second, each cycle requires 1/4 second. Thus the **period** of the oscillation is ¼ second. So we see that frequency and period are inverses of each other.

The frequency of sound is primarily experienced by humans in terms of pitch. (However, loudness also has a slight effect upon perceived pitch.) The audible frequency range for a young adult with normal hearing extends from 20 Hz to 20,000 Hz. At the lower end of this range, it is difficult to determine when the sensation passes from feeling to hearing, and there is much variability as to the upper end of the range also, depending upon the person's age, size, gender, and previous noise exposure. However, this range encompasses the commonly accepted range of audible frequencies. Data showing the average hearing loss *vs.* frequency with age in the developed nations indicate that sensitivity losses up to 45 dB are common in people aged 60-69, while more modest losses appear in populations in their 30's. (Tests of elderly people in less-developed nations seem to indicate that hearing does not naturally decline with age so much as with exposure to loud sounds: some octogenarians have been found to have hearing comparable to that of an American teenager.)

Table 2-A gives the frequency range of several common sound sources.

Table 2-A: Frequency Ranges of Common Sound Sources

Source	Frequency Range (Hz)
Pipe organ	16 to 8000
Piano	16 to 5274
Bassoon	29 to 415
Bass Tuba	36 to 349
Bass Voice	51 to 349
B-flat Trumpet	164 to 1046
Guitar	82 to 659

Soprano Voice	233 to 1046
Flute	246 to 1174
Piccolo	587 to 4186

In looking at this table, be aware that the sound wave from a real source is almost never a pure sine wave: the graph would show a different shape, as illustrated in Figure 2-2. These waves are made up of a fundamental frequency plus a number of frequencies that are integer multiples of that frequency. These higher frequencies are called **harmonics**. So although a piano can create fundamental frequencies extending from 16.351 to 5274.04 Hz, it can also create harmonics whose frequencies extend beyond the range of hearing. The same is true for many other sources.

In considering frequency, it is often helpful to correlate frequencies of sound with musical octaves. When the pitch of a note is increased by an octave, the frequency is doubled. Thus the frequency of middle C is 261.625 Hz; C an octave higher has a frequency of 523.25 Hz; and C an octave below middle C has a frequency of 130.81 Hz.

Now let us suppose that our vibrating object, perhaps a drum head or the top of a violin or guitar, is in contact with air. As the object vibrates, it strikes air molecules and causes them to vibrate as well. The graph of displacement *vs.* time for the air will look the same as the graph for our vibrating object. As the air is pressed outward by the vibrating object, there will be an instantaneous increase in the local density of the air, or **compression**. Just as increasing the density of air in a balloon by blowing into it creates pressure, so the object striking nearby air molecules causes an instantaneous pressure increase. As the compressed air naturally returns to its normal density, it passes along the compression to neighboring air molecules, and a pressure disturbance moves through the air. Likewise, as the object moves inward, air is drawn in behind it, decreasing the local density, or **rarefying** the air. The total effect of the vibrating object's contact with air is a succession of compressions or high-pressure regions, interspersed with rarefactions, or low-pressure regions that move away from the vibrating object. This is called an **acoustic wave**. If the wave happens to have a frequency that is within the range of human hearing, it is a **sound wave**. The spherical surface of air molecules that have been disturbed by the vibrating object is called the **wave front** of the sound wave. If we graph the instantaneous pressure in our acoustic wave *vs.* time, the graph will look just like the graph of displacement of the vibrating object that caused the acoustic wave.

The strength of an acoustic wave can be expressed in several ways. The most common way is the effective pressure. This is the time-averaged difference between the normal static pressure of the atmosphere, without respect to whether the instantaneous pressure is greater or less than the static pressure. Pressure is just force per unit area, so in the Imperial system, it is measured in pounds per square inch. In the SI system, pressure is measured in pascals, which are newtons of force per square meter.

Because much of the emphasis in acoustics is upon sound that is heard by humans, it is convenient to express sound pressure in terms that accord with human perception. Humans perceive most stimuli, including sound and light, in a more or less logarithmic fashion; *i.e.*, a sound wave whose effective pressure is doubled will not seem twice as loud, or a light whose intensity is doubled will not seem twice as bright. Instead, the sound or light will seem to have increased by about 0.3 times (the base-10 logarithm of 2), or 30%. Thus it is convenient to represent effective sound pressure in a logarithmically-related way. Initially, this was done by expressing the effective level in bels: the effective pressure in bels is the logarithm of the ratio of the effective pressure being measured to the effective pressure at the threshold of hearing. However, bels were found to be inconveniently large, so usage was changed to tenths of bels, or **deci**bels. Effective sound pressure expressed in decibels referenced to the threshold of hearing is called **sound pressure level**. Thus the sound pressure level at the threshold of hearing is defined as zero dB$_{SPL}$, since the ratio of effective pressures is unity, and the logarithm of one is zero. Sound pressure level in dB is 20 times the logarithm of the pressure ratio. So in the case mentioned above, in which the pressure increases by a factor of 2, the dB$_{SPL}$ change is 20 X 0.3 = 6dB. In order to seem twice as loud, a sound must increase in level by 10 dB, a pressure change of 3.16 times.

Sound pressure levels corresponding to several common environments are given in Table 1-B.

Table 2-B: Sound Pressure Levels (After Cavanaugh, p. 9 and Everest, p. 32)

Source	Sound Pressure Level	Sensation
Jet engine at 75 ft.	140 dB	Painful
Jet takeoff at 300 ft., max. rock band	120 dB	
Pneumatic chipper at 1.5 meters	110 dB	Deafening
Popular music group	95 - 100 dB	Very loud
Heavy truck passing by at 15 meters	85 dB	
Motion picture theater, average street traffic	80 dB	Loud
Conversational speech	65-70 dB	
Active business office	60 dB	
Private Office	40 dB	Moderate
Quiet living room	35 dB	Faint
Quiet concert hall	25 dB	
Rustle of leaves	15 dB	Very faint
Threshold of hearing	0 dB	

The strength of an acoustic wave can also be expressed as **acoustic intensity**, which is power per unit area. Acoustic intensity can also be expressed in decibels, and is given as 10 times the logarithm of the intensity ratio. So changing the acoustic intensity by a factor of 2 amounts to a 10 X 0.3 = 3 dB change in intensity.

When an acoustic wave moves through some material medium, it has a constant speed depending upon the mass density and stiffness of the medium. Since matter generally expands when heated, the mass density, and therefore the speed of sound in the medium, is temperature-dependent. In air, the mass density also depends upon the humidity. The speed of sound in various materials is given in Table 2-C.

Table 2-C: Speed of Sound in Various Materials

Material	Speed of Sound (meters/second)
Air (dry, 0°C, 760 mm Hg)	331.45
Water (distilled)	198
Aluminum (rolled)	6420
Copper (rolled)	3750
Silver	2680
Steel, mild	5200
Glass, light borate crown	4540
Oak, along the fiber	3850

As a sound wave moves through a material, a given point on the wave moves a certain distance during each cycle. This distance is called the **wavelength**. Wavelength is given by the speed of sound in the material divided by the frequency. Thus a middle C, having a frequency of 261.625 Hz, has a wavelength in dry air at 0°C of $\dfrac{331.45 \text{ meters/second}}{261.625 \text{ Hz}} = 1.267 \text{ meters}$. Many factors in acoustics depend upon the relation of physical size to the wavelength of the sound.

If a sound is created by a small object such as a small balloon bursting, the sound waves will move outward in all directions from the center of the burst. This is called a **spherical wave**. As the wave expands, the total surface area of the sphere representing the initial wave front expands proportionally to the square of the distance from the source, since the area of a sphere depends upon the square of the radius. As we are ignoring the effects of friction, and there is no way that additional energy can be added to the sound wave as it progresses; the intensity, and therefore, the sound pressure level, must decrease in proportion to the increase in surface area: **sound pressure level in an open space decreases proportionally to the square of the distance from the source**. This **inverse square law** is also a property of light and of other waves in three-dimensional space. Considered in decibels, this translates to: **sound pressure level in an open space decreases by 6 dB with each doubling of distance from the source**.

If a sound source is not small compared to the wavelength of the sound, it will in general not form a spherical wave, but will travel preferentially in one direction, depending upon the shape of the source. The human voice and a trumpet are two common directional sources. The wave front will be represented by a section of a spherical surface. The inverse square law applies to directional sources as well.

If the air through which a sound wave passes is uniform, the wave front will move in a straight line outward from the source. If the wave encounters an object whose density and/or stiffness is different from those of air, part of the wave will be transmitted to the object, and part will be reflected back. The reflected portion of the wave will obey the Law of Reflection: the angle formed by the perpendicular to the incident wave front and the perpendicular to the surface of the object (**angle of incidence**) equals the angle formed by the perpendicular to the reflected wave front and the perpendicular to the object (**angle of reflection**). The Law of Reflection is illustrated in Figure 2-3.

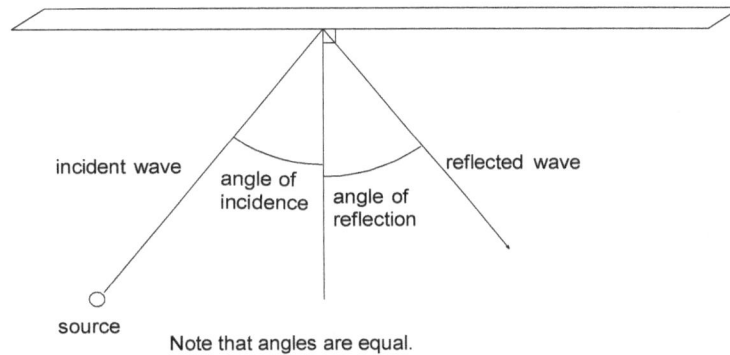

incident wave

angle of incidence

angle of reflection

reflected wave

source

Note that angles are equal.

Figure 2-3: The Law of Reflection

The portion of the sound wave that is transmitted into the object may proceed in that object with a different direction from that of the original incident wave. An example of this effect occurs in very large outdoor concerts, where changes in air temperature with elevation above the ground cause the speed of sound to change. The result is that the sound from the stage can curve up; missing the rear of the audience, or it may curve down, making the sound in the rear unusually loud. This curving of the path of a sound wave is called **refraction**, and is analogous to the refraction of light waves by eyeglasses. In the case of eyeglasses, though, the refraction is caused by the wave passing from one medium to another in which the wave travels at a different speed. Refraction of sound in air occurs because of a change in the properties of the air, resulting in a change in the speed of sound in the air as the wave progresses.

When a sound wave encounters an object much larger than a wavelength, a sound shadow will be formed behind the object. But if the object is much smaller than a wavelength, the sound will seem to bend around the object. This phenomenon, called **diffraction**, is the reason that when recorded music is played in another room, you hear mainly the low bass frequencies. The obstacles – walls, furniture,

19

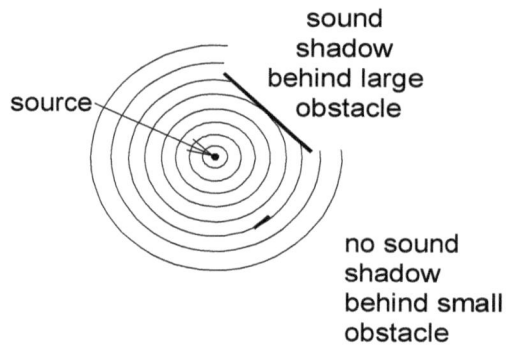

Figure 2-4: Diffraction

etc. – are large compared to the short wavelengths of the high frequencies, but small compared to the long wavelengths of the low frequencies. Thus the low frequencies diffract around the obstacles that block the high frequencies. Diffraction is illustrated in Figure 2-4.

Both when discussing directional radiation of sound, and when discussing diffraction, reference has been made to whether an object is large or small compared to a wavelength. The question naturally arises, just how large or small must an object be to qualify as "large" or small"? In general, any vibrating object that is larger than ¼ wavelength across can have directional effects, and any obstacle larger than ¼ wavelength across can create a sound shadow. Thus, recalling our earlier example, a speaker would have to be at least 1.267m/4 or about 32 cm to have any directionality at the frequency of middle C, and likewise an obstacle would have to be larger than about 32 cm across to shadow sound at that frequency.

3. Physiological Acoustics
Loudness

Figure 3-1 shows a schematic drawing of the human hearing mechanism. The outer ear or **pinna**, which we usually just call "the ear," serves to help pick up sound vibrations to be transferred into the middle and inner ear. It also changes the incoming sound wave's frequency spectrum according to the direction from which the sound arrives. These changes are responsible for part of our ability to localize sound sources in space.

Sound travels from the outer ear through the auditory canal to the eardrum, a thin membrane that terminates the auditory canal. The resonant action of the auditory

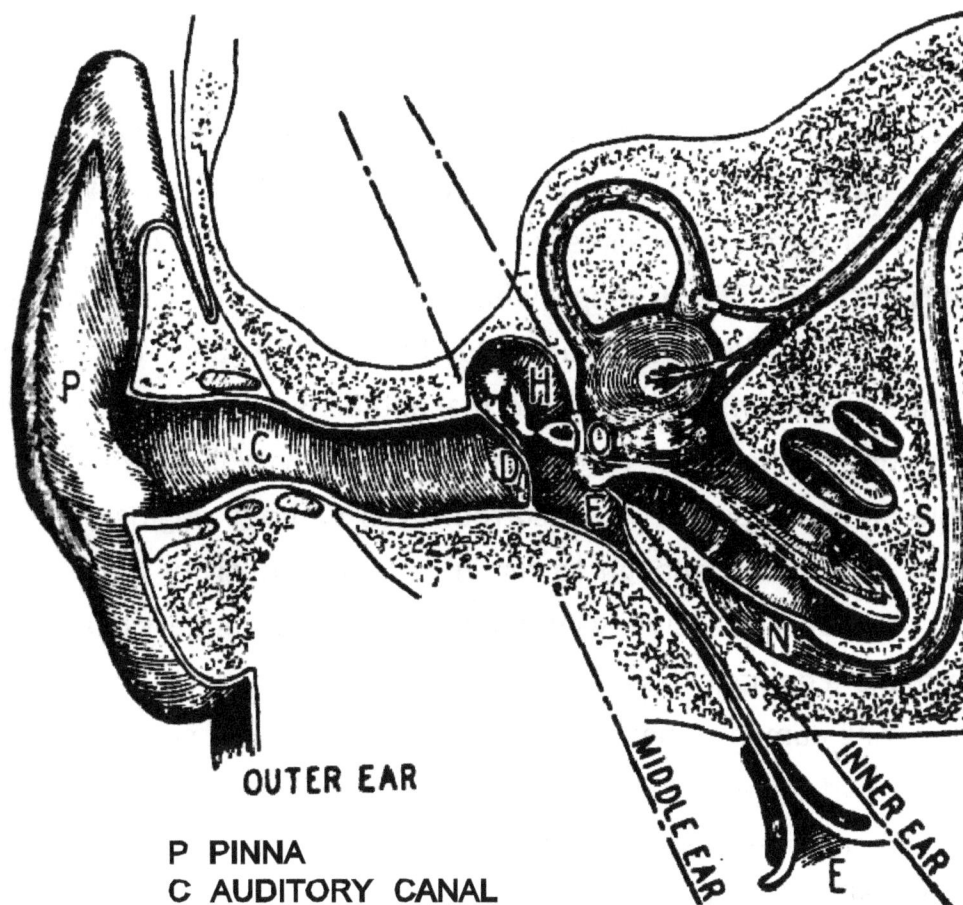

OUTER EAR

P PINNA
C AUDITORY CANAL

Figure 3-1: The Human Ear (Knudsen, p. 73)

canal intensifies sound in the range of 3000 - 4000 Hz by about 5 dB. These frequencies are very important in understanding speech. Attached to the back of the eardrum is the **malleus**, the first of three bones called **ossicles**: "hammer", "anvil", and "stirrup". Together with the eardrum, these bones translate the relatively large-motion, low-force sound wave in air into a smaller-motion, higher-force wave in the fluid of the cochlea, to which the stirrup is connected. ("Large-motion" must be taken as a comparative term: the actual motion of the eardrum at the threshold of hearing is less than the diameter of a hydrogen molecule!) The ossicles may be compared to a system of levers.

After the vibrations enter the cochlea, they are converted to motions in tiny hair cells or **stereocilia**, and are thereby transformed into nerve impulses which are sent to the brain, which extracts information on loudness, pitch, direction, etc.

The most common cause of hearing loss is damage to the hair cells, which is irreversible, although some research is being done with a view toward our being able to repair or regenerate hair cells in the future. Not surprisingly, the first range to be lost is the 3000-4000-Hz range, where the ear canal intensifies the sound. The cause of most hearing loss is exposure to loud sound. The dangerous misconception persists that "ringing in the ears" is normal and all is well once it goes away. The fact is that an experience of ringing in the ears may well be an indication that irreparable damage has already occurred. Even without ringing, if the ears are exposed to sound loud enough to diminish their sensitivity (hearing threshold shift), and the reduced sensitivity lasts over 24 hours (secondary hearing threshold shift), permanent damage has likely occurred. Much has been written about noise-induced hearing damage from loud musical groups – even symphony orchestras. Less has been written about the damage from chain saw noise, snowmobile noise, and other "recreational" sources of noise. In order to protect our hearing, we should not expose our ears to noises greater than about 80 dB for long periods of time. If you can hum to yourself and cannot hear the humming inside your head over the external sound, the external sound is greater than about 80 dB. The louder the sound to which the ears are exposed, the briefer must be the exposure, if damage is to be avoided. There are tiny muscles that restrict the motion of the ossicles when the ears are exposed to loud sound, but these muscles tire and cannot do their job if the exposure is too long.

The other common cause of hearing loss is immobilization of the ossicles, which causes a loss of low-frequency sensitivity. This problem can often be corrected by a surgery called a "stapes mobilization."

The sensitivity of the ear has long been known to depend heavily on the frequency of the sound. In 1933, Fletcher and Munson of the Bell Telephone Laboratories published a set of curves showing the sensitivity of the ear *vs.* frequency. These curves have been updated by Robinson and Dadson and adopted as I. S. O. Standard 266, and are shown in Figure 3-2. Notice that the bottom curve represents the threshold of hearing, which varies from about 76 dB for a frequency of 20 Hz to about -3 dB (3 dB below what was formerly considered the "threshold of hearing") at 4000 Hz, to about 22 dB at the 25,000-Hz limit of measurement. However, not only does sensitivity vary with frequency, but the range of variation changes with level. While, at the threshold of hearing, there is a 79-dB difference between the 20-Hz sensitivity and the 4000-Hz sensitivity; when the sound level is about 100 dB, the difference between the 20-Hz and 4000-Hz sensitivities is only about 42 dB. This is the reason that music played softly seems deficient in bass compared to louder music. In order to take the ear's relative insensitivity into account, sound levels are sometimes measured with "A weighting", a method that de-emphasizes the contribution of low frequencies to the sound pressure level. Such measurements are specified in **dBA**.

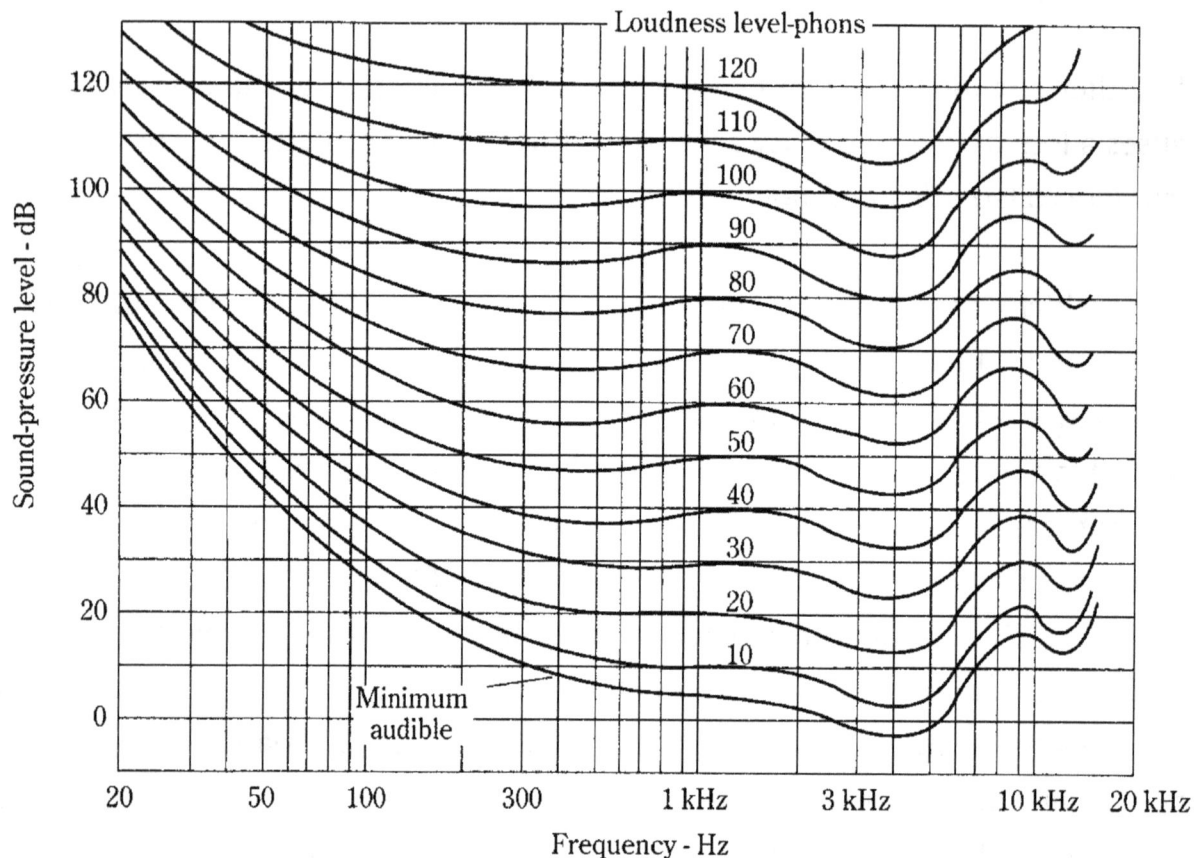

Figure 3-2: Equal-Loudness Contours (After Robinson and Dadson)

By now is should be clear that sound pressure level expressed in dB is a physical, not a perceptual, measurement. Much work has been done to establish a measure of loudness that corresponds to human perception. One great stride was made with the introduction of the **phon** as the unit of loudness level. A loudness level of 40 phons is defined as the loudness experienced by a human observer listening to a 1000-Hz sound at 40 dBSPL. This is the same loudness experienced when the listener hears a 20-Hz tone at 92 dBSPL, as can be seen from Figure 2-2. Thus the phon scale takes into account the effect of frequency on loudness.

However, the phon scale does not correspond with the human perception of changes in loudness. In chapter 2, the fact was mentioned that humans experience changes in sound pressure level logarithmically. In fact, if the sound pressure level of a sound is increased so that it sounds twice as loud as before, the necessary increase is about 10 dB. In order to overcome this difficulty, the **sone** scale was introduced. One sone is defined as the loudness corresponding to a loudness level of 40 phons. A loudness level of 50 phons corresponds to 2 sones; 60 phones, to 4 sones, etc. Thus when the perceived loudness doubles, so does the loudness expressed in sones.

A further complicating factor in human perception of loudness is bandwidth. If a complex sound contains various frequencies, all close together, it will not seem to be as loud as another sound having many frequencies spread farther apart, even though the sound pressure level in dB is the same.

Localization

The ability to locate the position of a sound source had critical importance to our ancestors. Since most hazards (and most potential prey) were located at ground level, they developed the ability to aurally localize with great accuracy (within a few degrees) in the horizontal plane. However, localization in the vertical plane is less precise.

The primary characteristic of a sound we use in localization is time of arrival. If we hear two identical sounds, our brains assume the one heard first is the original sound, while the second may be an echo. This fact is often called "the law of the first wave front." The brain can respond to tiny time differences corresponding to the arrival of a sound wave at the left and right ears.

The second characteristic used in localization is relative level. Especially for high frequencies, the sound will be louder at the ear pointed more directly toward the source.

The third characteristic is based upon the effect of the pinna upon the frequency-sensitivity of the ear. As mentioned earlier, the pinna produces changes in this sensitivity depending upon the direction of arrival of a sound.

The combination of these latter two effects are called a "head-related transfer function," or HRTF. By creating artificial HRTF-like effects through electronic manipulation of recorded sounds, directional effects can be mimicked.

Discrimination of auditory events

When we listen to sound in a room, we hear not only the original sound produced by the person speaking, singing, or playing music, but also many reflections from the room walls, ceiling, and floor. Only in extreme cases do we detect these reflected sounds as discrete auditory events, or echoes. Usually, our ear/brain combination blends them into a uniform whole. In 1951, Helmut Hass conducted experiments in which listeners were presented a direct sound through one loudspeaker, then a second, delayed sound through another loudspeaker. He asked the listeners to adjust the level of the delayed sound until they could clearly perceive an echo. Haas found that if the second sound occurred within 35 milliseconds (ms) of the original sound, it was not perceived as a separate sound, but that it only made the first sound seem louder, with perhaps an increase in the apparent (acoustical) width of the source. This was true even when the delayed sound was up to 10 dB louder than the original. Beyond 50 milliseconds, the delayed sound was perceived as an echo with the threshold of perception as an echo becoming lower and lower as the time delay was increased. Some authorities consider the "Hass interval", or "integration interval" to be 35 ms, and others consider it to be 50 ms. Subsequent research has shown, not surprisingly, that the "integration interval" during which the delayed sound is perceptually fused with the original depends upon the program material. Sharp rim shots on a snare drum permit the delayed sound to be heard separately with a much shorter delay than do extended violin passages. The frequency content of the signal also affects the integration interval: for sounds around 500 Hz (octave above middle C), the interval is about 45 ms; whereas, for sounds two octaves higher (around 2000 Hz), the interval is closer to 25 ms. One can better relate to these intervals if one realizes that sound travels about 1 foot (0.3 meters) per ms. Thus a 25 ms delay corresponds to a reflection traveling about 25 feet (7.5 meters) distant; indicating a reflecting surface 12.5 feet away; 45 ms corresponds to 45 feet (13.5 meters) of travel, or a reflector 22.5 feet (6.75 meters) away.

Masking

Everyone is familiar with the fact that a loud sound can prevent one from hearing a softer sound. This is the phenomenon of **masking**. But masking is much more complex than it seems on the surface. When someone says the word "sound", certainly we hear an acoustical event, a sound. But what we hear can also be described as a sequence of sound waves of varying frequency and intensity. First, there is the unvoiced "s", composed of high frequencies and relatively low intensity. Next there is the "ou", comprising lower-frequency waves of relatively higher intensity. Then finally, the voiced "n" and "d", mainly consisting of low frequencies at an intermediate intensity level.

Masking has been found to depend not only upon intensity (loud sound masks soft sound), but also upon frequency. The closer in frequency two sounds are, the more likely it is that masking will occur. This is why a fairly low level of "whish" from an HVAC system can mask the crucial consonants (especially the unvoiced ones) in speech. Thus in a noisy room, one can only hear "the ound o mu-i", not the full sound of music.

Masking is not strictly a function of physical hearing, but is an ear-brain function. With focused attention, it is possible to understand words in conditions that would be expected to mask speech. An example of this is the "cocktail-party effect", by which one can understand a conversation on the other side of a party room, even though the sound level of the conversation is below that of the ambient noise. Hearing-aid users are not able to discriminate among sounds arriving from different directions as well as those having normal hearing, and some hearing aids compress the dynamic range (loudest-to-softest ratio) of the sound. These factors cause hearing-aid users to experience more speech masking than do those with normal hearing.

4. The Nature of Speech, Music, and Noise

Speech, music, and noise can be characterized by purpose: speech is used primarily for communication; music, primarily for emotional and/or aesthetic effect; and noise is defined as unwanted sound.

Speech and music involve different sound levels, ranges of frequencies, and time characteristics. Figure 4-1 shows the average levels and frequency ranges in speech for men and women.

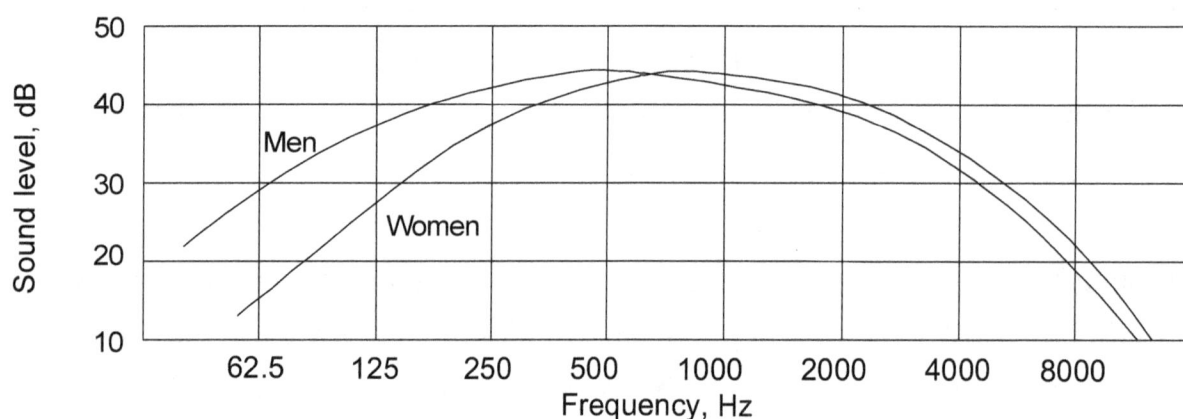

Figure 4-1: Frequency/Level in Male and Female Speech (after Fletcher)

From this figure, we can see that most of the energy in male and female speech is concentrated about 600 Hz, although the frequency extremes extend from about 40 Hz to 9,000 Hz.

Figure 4-2 provides more detail in that it illustrates the frequencies and levels of vowels, voiced consonants, unvoiced consonants, and sibilance, as well as where these vocal features fall in the hearing spectrum between the threshold of hearing and the threshold of feeling.

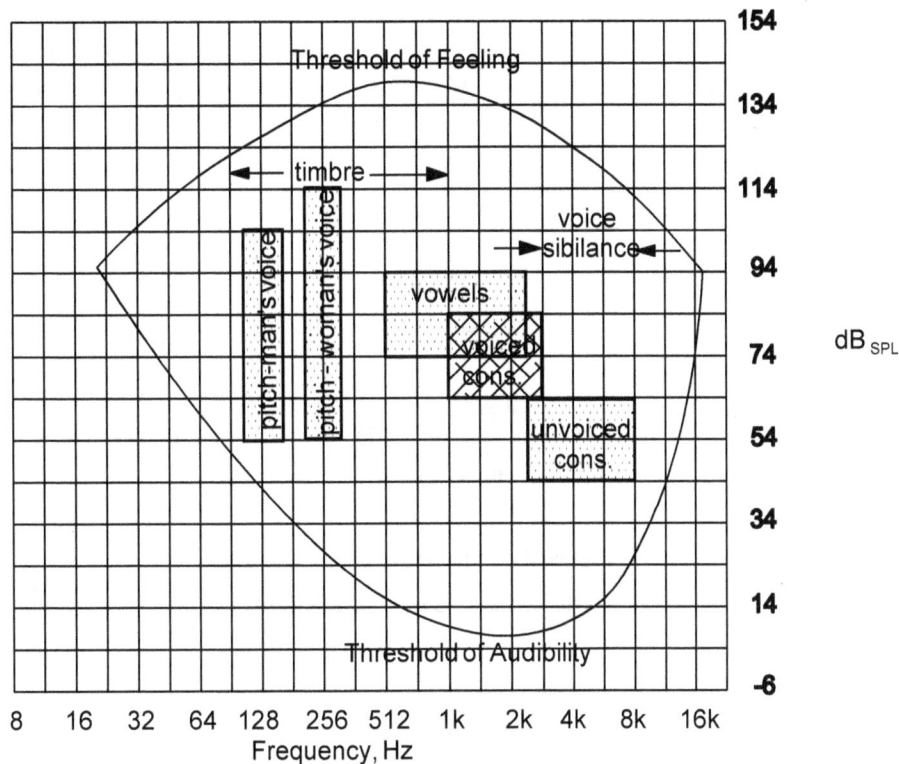

Figure 4-2: Speech Frequencies and Levels
(After Steinberg, in Knudsen)

Looking at this figure, we can easily understand why people experiencing a hearing loss at high frequencies have difficulty understanding speech: the consonants, and particularly the sibilants, may be inaudible to them. Likewise, rooms that acoustically reinforce low-frequency sounds more than high-frequency sounds will prove problematic for the understanding of speech.

Figures 4-3 through 4-6 show the frequencies and levels produced by a bass drum, a piccolo, and a 75-piece orchestra.

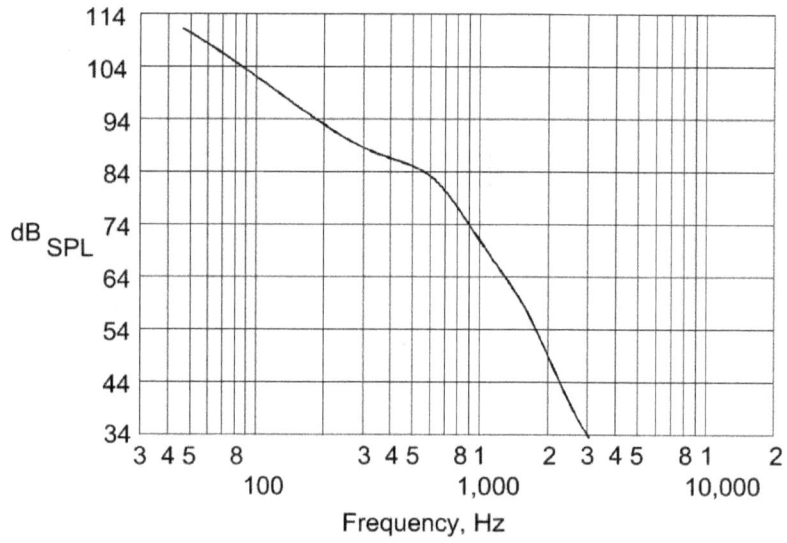

Figure 4-3: Spectrum of Bass Drum

(After Sivian, Dunn, and White)

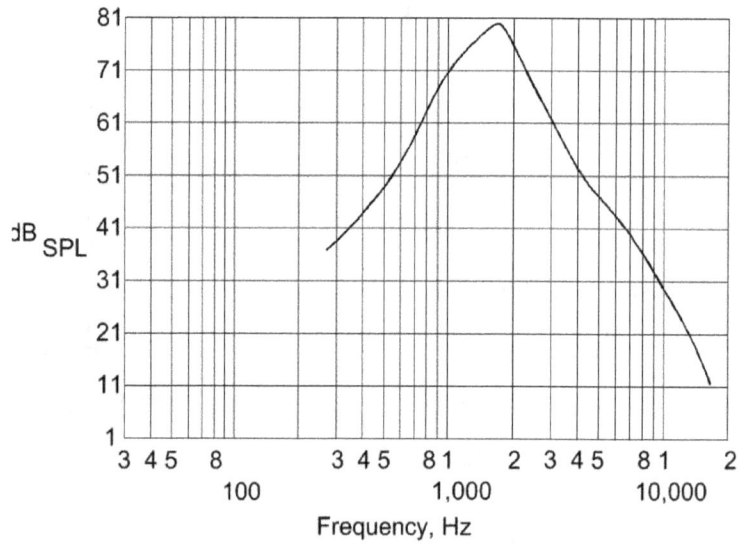

Figure 4-4: Spectrum of Piccolo

(After Sivian, Dunn, and White

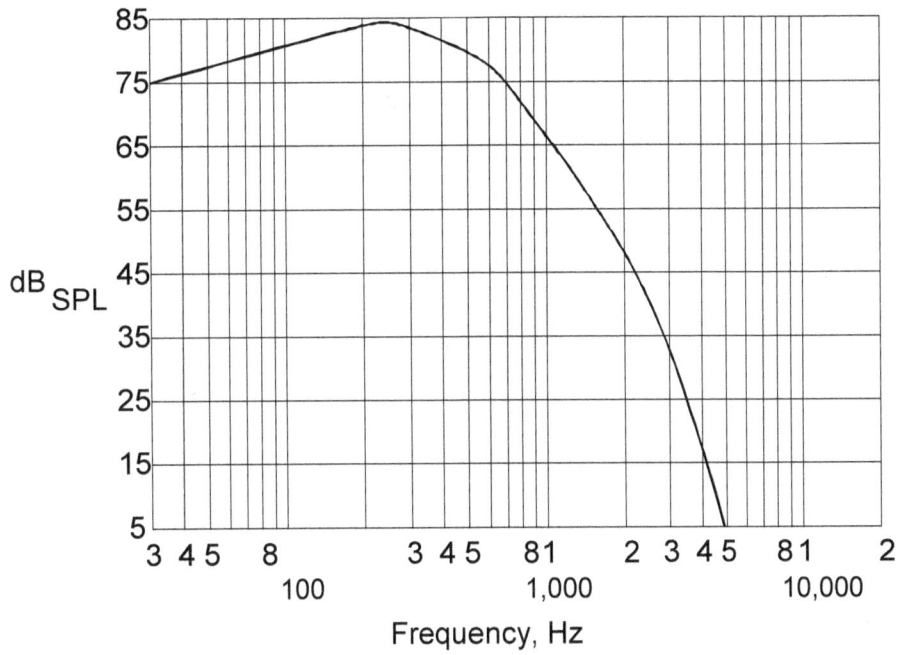

Figure 4-5: Spectrum of Piano
(After Sivian, Dunn, and White)

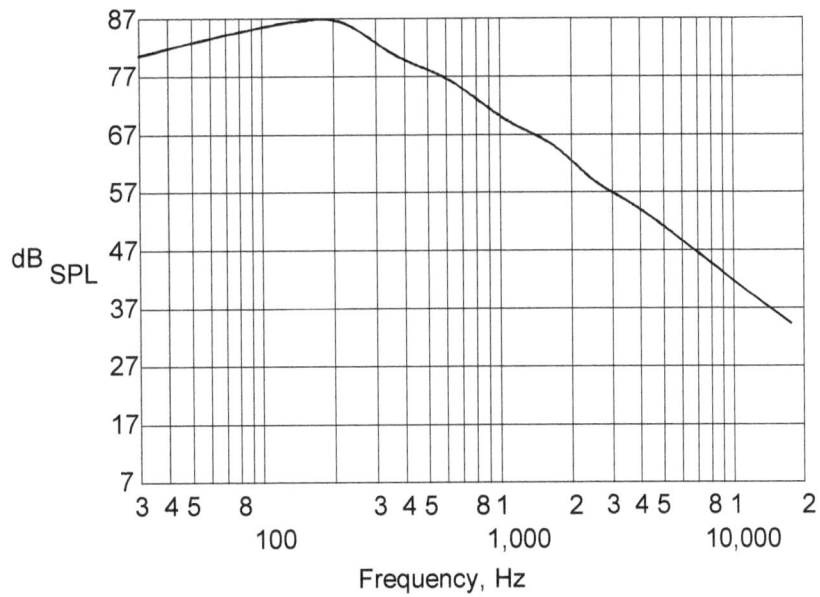

Figure 4-6: Spectrum of 75-Piece Orchestra
(After Sivian, Dunn, and White)

Notice that both the sound level and the frequency range produced by unamplified musical instruments exceed those produced in speech. Whereas speech encompasses a range of about 40 – 9,000 Hz, music can cover at least 20 – 17,000 Hz. Some musical instruments can produce sounds in excess of 20,000 Hz, although tests at the BBC indicate that few people are aware of the presence or absence of such high-frequency components in music. (Shorter, Manson, and Wigan)

Any source of sound will radiate more favorably in some directions, and less favorably in others, and this directivity depends upon frequency. The human voice and small musical instruments radiate nearly equally in all directions at low frequencies, but become directional when the size of the radiating body approaches the wavelength of the sound. Thus we would expect that chest tones from a male singer would become directional about 550 Hz (roughly C# an octave above middle C), where the wavelength is about two feet. Of course, at this frequency, chest tones are weak, so the voice is relatively omnidirectional (radiating equally in all directions) for chest tones. At increasingly higher frequencies, the voice radiates preferentially forward and slightly downward.

The directivity of orchestral instruments is more complex. Wind instruments behave similarly to the voice, but stringed instruments radiate in complicated ways, with the high frequencies coming off mainly perpendicular to the top plate, and low frequencies being omnidirectional. High frequencies from a piano are affected by the position of the case lid: closed, open on short stick, or open on long stick. Organs are essentially omnidirectional.

Noise is defined as undesired sound. Noise as an acoustical problem may originate from people or machinery or other devices outside the space under consideration, or from internal sources such as water coolers, video projectors, or ventilation systems within the space. Noise can be characterized in terms of sound level and frequency range. Low-frequency, "rumbling" noise is often difficult to eliminate; whereas high-frequency, "hiss" noise yields more easily to abatement efforts. Besides the sound level, two characteristics of noise can make it especially annoying. Noise that contains intelligible information significantly interferes with concentration and thus is annoying. Presumably this is because the brain attempts to decipher the intelligence in the noise. Also, noise containing pure tones (whistles or hums with determinable pitches) is quite annoying. Part of the reason for this is that humans are capable of discerning pure tones buried in noise that has a significantly higher sound level than the tone.

Acoustic noise is not always bad: sometimes random noise is used to mask speech in adjacent offices, and some people find recordings of seaside sounds to be relaxing. (Of course, the question arises whether these sounds, being "desired", are noise after all!)

II Physical Acoustics of Rooms

5. Reverberation

Reverberation is the sound that bounces around in a room after the initial sound. The time required for a sound at approximately the level of ordinary conversation to fade to inaudibility in a very quiet space is denoted as the *reverberation time*, symbolized as RT60. Figure 5-1 shows the sound decay curve for a fairly dry room.

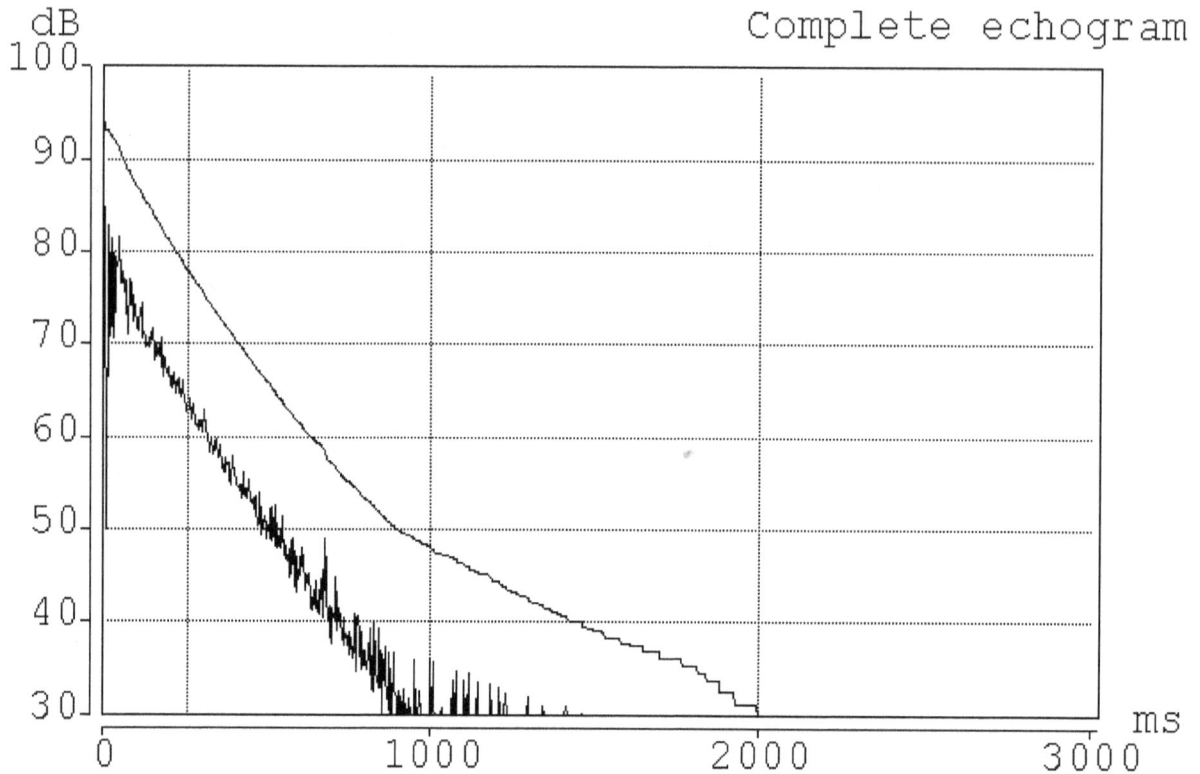

Figure 5.1: Sound Decay Curve

The jagged line shows the many individual reflections that, together, make up the reverberation. The smooth line is sort of a running average produced by a mathematical process called *Schroeder backward integration*. The original definition of reverberation time would be the time required for the smooth line to drop from its beginning – at a level of about 94 dB – by 60 dB. In this room, background noise skews the decay curve at levels below about 50 dB, so an alternative method is used: RT60 is considered to be twice the time required for the level to drop from 5dB to 35 dB below the maximum. The initial 5-dB is ignored, because initial reflections skew the decay curve. Thus the RT60 in this room is about twice the time between when the level is 5 dB below the maximum (about 57 ms) and when it is 35 dB below the maximum (about 632 ms), giving RT60= 1.15 seconds.

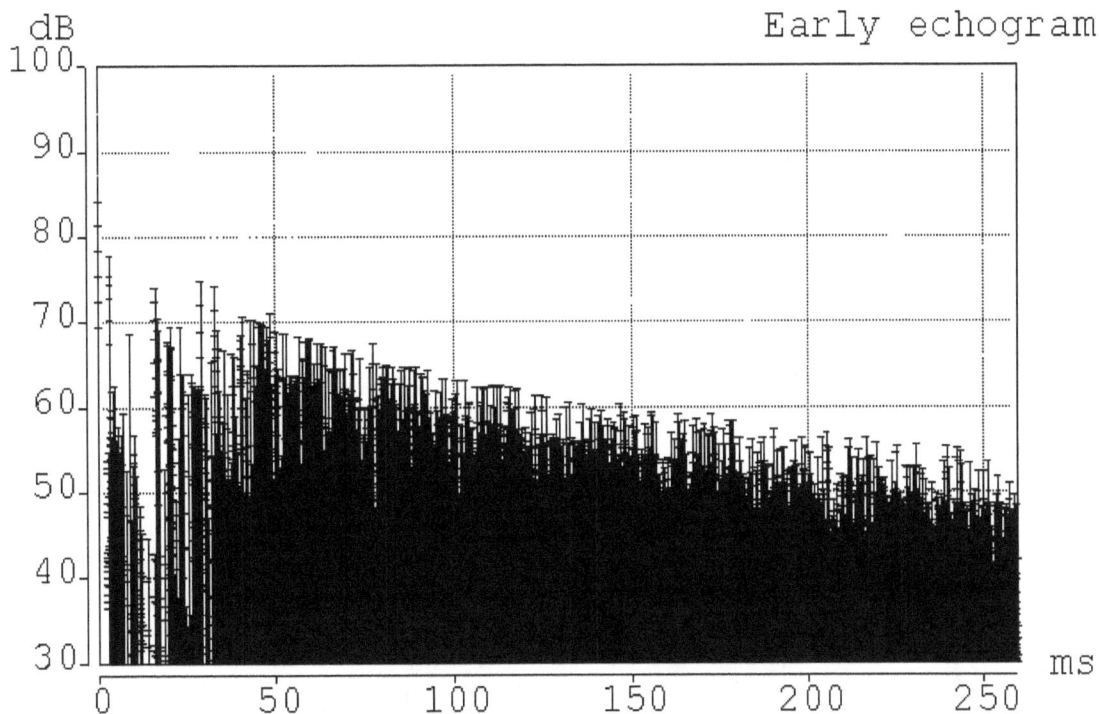

Figure 5-2: Early Sound Decay

Figure 5-2 shows the first ¼-second of the decay curve for the same room. The horizontal "hash marks" on the lines representing reflections indicate how many reflections occur at the same point in time. Notice that the original sound plus reflections from very near the sound source occurs (left-hand side of the graph), and then there is a very short time before the sound has traveled to an object and bounced back off it. Then there are a few reflections that stand out by themselves, followed by a very crowded group of reflections making up the "reverberation tail". Those initial separated reflections are the first ones from the walls, ceiling, and floor, and they represent the way in which our ear-brain system allows us to perceive the size of the room we're in. In other words, they are much of the way in which we can aurally distinguish between a reverberant bathroom and a reverberant auditorium.

Another measure of reverberation that is very important for music rooms is the **early decay time** (EDT). This is 6 times the time required for the sound to decay by 10 dB from its initial maximum. Must of people's judgment of the reverberance of a room is determined by EDT. In many rooms, EDT is very nearly the same as RT60, but there are architectural features that can cause differences between the two.

After a sound source has been active in a room for some time, the time-averaged level of the reverberant sound is the same everywhere in the room, although the sound heard directly from the source decreases with distance from the source. This state of affairs is illustrated by Figure 5-3.

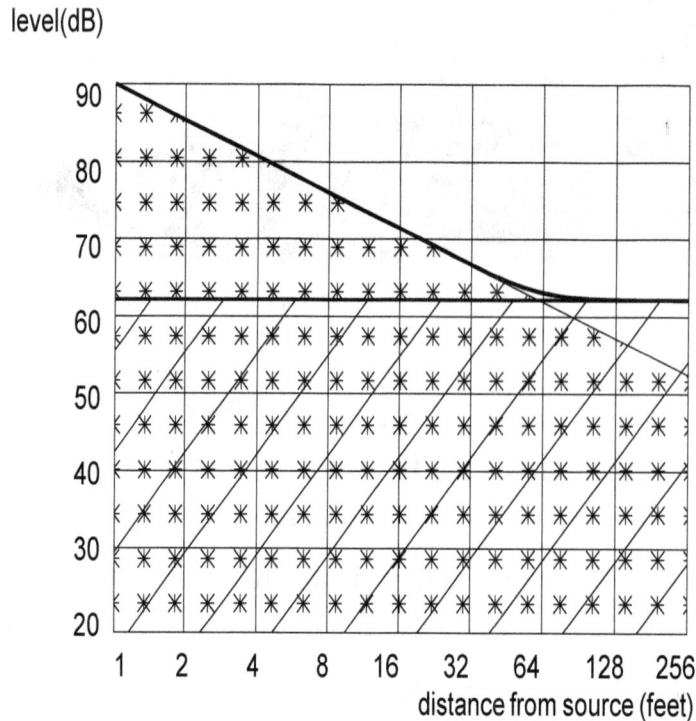

Figure 5-3: Sound Level Variation with Distamce in a Room

The line sloping down to the right illustrates the sound level heard directly from the source – the so-called "direct" sound. The area under this line is hatched with a star pattern. The horizontal line at 62 dB shows the level of reverberant sound, which in this case could be typical for a commercial cafeteria. Notice that at a distance of 64 feet from the source, the direct and reverberant sound levels are equal, and the heavy curved line shows the total sound level heard by a person at various distances from the sound source. There is a special name for the distance where the direct sound and reverberant sound have the same level: the **critical distance**. Of course, this analysis ignores the effect of background noise, so the distance at which direct sound begins to be swamped by reverberation and noise is always somewhat less than the critical distance.

What is the optimum RT60? What is the optimum weight for an athlete? Both questions have the same answer: it depends. Speech intelligibility benefits from having RT60 as short as possible. The optimum RT60 for music depends heavily upon the type of music: organ music benefits from a long RT60; chamber music, from a shorter one, and modern amplified music, from a very short reverberation time. It is not only the musical style that affects the optimum reverberation time. The rapidity of movement of the music is also a very important consideration. While slow solemn organ music sounds very good in a room having a 3-second RT60, a piece of modern French music – or even Bach – played at an allegro or faster tempo or abounding in short, rapid notes will sound smeared and confused in a room with that much reverberation. Thus a room designed to support traditional-hymn-style organ music, with tempi usually between 60 and 100 beats per minute and melody lines composed primarily of quarter and eighth notes, probably should have a reverb time of about 1.8-2 seconds; whereas, a room designed to support the long, slow notes of Gregorian chant should have a longer RT60.

The musical effect of reverberation is well-known by most people: it prolongs and connects the sound. The effect on speech is to reduce the intelligibility of the speech. Thus in a room used for both speech and music, such as a multipurpose auditorium or church sanctuary, a compromise is often necessary between the optimum RT60 for speech, and the optimum for the particular musical style. This is less true for musical styles involving fast music and/or prominent use of percussion instruments, since such music sounds best with a relatively short RT60.

Table 5-A shows the range of reverberation times preferred for rooms designed for a variety of functions, along with the typical range of cubic volume of such rooms. (The preferred RT60 is higher for larger rooms.)

Function	Range of RT60	Range of Typical Volume (cu. ft.)
Lecture. meetings, drama	<1.1 s	<140,000
Multipurpose halls, music & drama, sports	1.25 – 1.5 s	150,000 – 800,000
Opera, chamber music	1.5 – 1.7 s	250,000 – 700,000
Symphonic Music	1.75 – 2 s	350,000 – 900,000
Organ, oratorio	2.1 – 2.5 s*	400,000- 1,000,000

Table 5A: Preferred Reverberation Times

* Duke University Chapel in Durham, NC has an RT60 of almost 7 seconds – suitable for a Gothic Cathedral, but useful only for specific types of music. Adequate speech intelligibility requires exceptional care and a large budget for sound-system design

38

The reverberation time of a room depends upon two physical factors: the cubic volume and the total absorption of the room. Volume comes into play because in a larger room, the sound travels farther between bounces, and since significant energy is only absorbed when the sound bounces off an object, fewer bounces per second mean that more seconds are required to absorb enough energy to reduce the sound level by 60 dB. The total absorption is controlled by the average absorptivity of the room surfaces and the total area of those surfaces. Thus a room having a surface area of 1000 square meters, and an absorptivity of 50% (meaning that 50% of the sound striking the surfaces is absorbed, and the other 50% is reflected back into the room), will have a total absorption of 500 *metric sabins*. (The *sabin* is the unit of absorption, and can be expressed in units of square feet or square meters.) If one wall has an absorption of 10%; another, 20%, and the other two, 30% and 40%, respectively, the average absorption is the average of the products of the areas and their respective absorptivity: the sum (10% of wall 1's area + 20% of wall 2's area + 30% of wall 3's area + 40% of wall 4's area) – divided by the total area. So if each wall has an area of 50 square meters, the average absorptivity of the walls becomes (5+10+15+20)/ (4X50) =25%. Notice that a small area of very absorptive material will make little difference in the average absorptivity; and hence, little difference in the RT60. The absorptive effect of ceilings and floors is calculated in the same way.

The absorptivity of a material depends upon the frequency of the sound. In general, thick materials are required in order to absorb low-frequency sounds well. For example, a 1" thick fiberglass panel absorbs almost all the sound striking it at frequencies above about 1000 Hz (roughly two octaves above middle C), but only about 30% of the sound at middle C (256 Hz). To get good absorption all the way down to middle C, we have to use 4" thick fiberglass. In the same way, we find that ordinary carpets, being thin, absorb mainly high frequencies.

The physical mounting of a material affects the sound absorptivity as well. If a 1" fiberglass panel is mounted over a 2" airspace, the effect is nearly the same as that of a 2" thick panel. This is why good acoustical ceilings absorb low frequencies reasonably well, even though the ceiling panels are seldom over 1" thick.

6. **Echoes**

Although both echoes and reverberation result from sound reflections, there is a difference between the two. Echoes are strong single reflections or groups of strong reflections occurring almost simultaneously, forming a distinctive acoustical event. Reverberation is a sequence of weakening reflections occurring over a time period ranging from a small fraction of a second to several seconds, but producing a perceived lengthening of an acoustical event rather than a distinct event. Thus an echo sounds like clap, clap, clap, while reverberation sounds more like cl-a-a-a-p. Reverberation can enhance the sound of music, but echoes are almost always detrimental, confusing the sound and decreasing speech intelligibility. Two conditions are necessary to create an echo: a relatively smooth, hard sound-reflecting surface, and sufficient distance from source to reflecting surface to listener that the echo is perceived as a separate sound. As was mentioned in chapter 3, the minimum time delay between original sound and echo needed for the echo to be perceived is usually given as the Haas interval of 35 – 50 milliseconds, which is the time required for sound to travel 40 – 57 feet. Thus a reflecting surface about 30 feet away from a person will cause an echo when the person claps, since the sound travels a round-trip distance of 60 feet.

The stronger an echo is compared to the original sound, the more likely it is to be heard as an echo. Weak echoes, or echoes occurring sooner after the original sound, are less easily noticed. Echoes are not always noticed only as discrete sound events; sometimes they confuse the ear's ability to localize a sound source; at other times, they can cause cancellation and/or reinforcement of sounds at various frequencies, making for an unnatural timbre. Echoes are more noticeable and more disturbing to speech than to music, and fast or percussive music is more sensitive to degradation by echoes than slow, legato music.

Echoes can be prevented by covering potentially reflecting walls with absorptive materials (4" fiberglass may be needed to prevent rear-wall echoes in auditoriums and movie theaters in order to provide enough absorption at low frequencies), or designing irregularities into the surface of the wall to diffuse or "break up" the reflections. Diffuse reflections travel in many directions; whereas, non-diffuse or specular reflections, like reflections of light from a mirror, bounce back from the reflecting surface in a single direction. Diffusion can be provided by acoustical treatments specifically designed for the purpose, or by architectural features. In many cases, a combination of absorption and diffusion is best.

7. Strength

Strength, also called "room gain", is the ability of a room's surfaces to reinforce the sound of a talker, singer, or musical instrument. This reinforcement occurs because of early reflections that occur within the Haas interval and therefore are merged by the ear/brain system into the same acoustical event as the original sound. An example of a structure designed to increase room strength is the orchestra or choir shell.

While less critical today, because of the common use of amplification systems, strength is still important for lecture halls, theaters for drama, and concert halls to be used for performance by quiet instruments such as the classical guitar.

There are some commonly-used features of auditoriums that decrease strength. Heavy stage curtains over the talker's head and behind the talker absorb sound rather than reflecting it to the audience. In fact, the way a shell helps strength is that to covers these absorbent surfaces with a reflecting one. Basically, any reflecting surface oriented so that sound from a talker or performer bounces from it to the audience, and located within 17-20 feet of the talker or performer, will increase strength. Reflective surfaces located farther away will increase the sound level compared to what one would have using absorptive surfaces, but the reflections from these more distant surfaces will not be perceived as part of the "program". Rather, they will be heard as early echoes, which may be detrimental to clarity and speech intelligibility. However, strength is also increased by reflective surfaces located near the listener. In these cases, the short time delay between direct and reflected sound comes about because of the short distance between the reflecting surface and the listener. An example is the use of formed reflective ceilings or acoustical clouds above the listeners. Even if the ceiling is not formed to enhance beneficial reflections, it should be reflective. Acoustical absorption should almost never be used on the ceiling of a performance or worship space. A major exception to this could be a venue for amplified contemporary music, especially if the volume is so large due to a very high ceiling that the reverberation would be high without absorption on the ceiling.

Sidewall reflections can increase strength, but only for those listeners seated fairly close to the walls.

Figure 7-1 shows the decay curve of a medium-sized church with the sound source in the choir loft. (Actually, three traces of the decay curve are shown: one illustrating the effect of reflections from horizontal surfaces, one showing reflections from side walls, and one showing reflections from end walls. Notice that there are three strong early reflections: one at 9 milliseconds and one at 26 milliseconds, both of which will add to strength, and one at 60 milliseconds that may be perceived as an echo. After these three early reflections, the sound level builds up for about 30 milliseconds as reverberation is established in the room, then decreases at a rate corresponding to an RT60 of about 1.2 seconds.

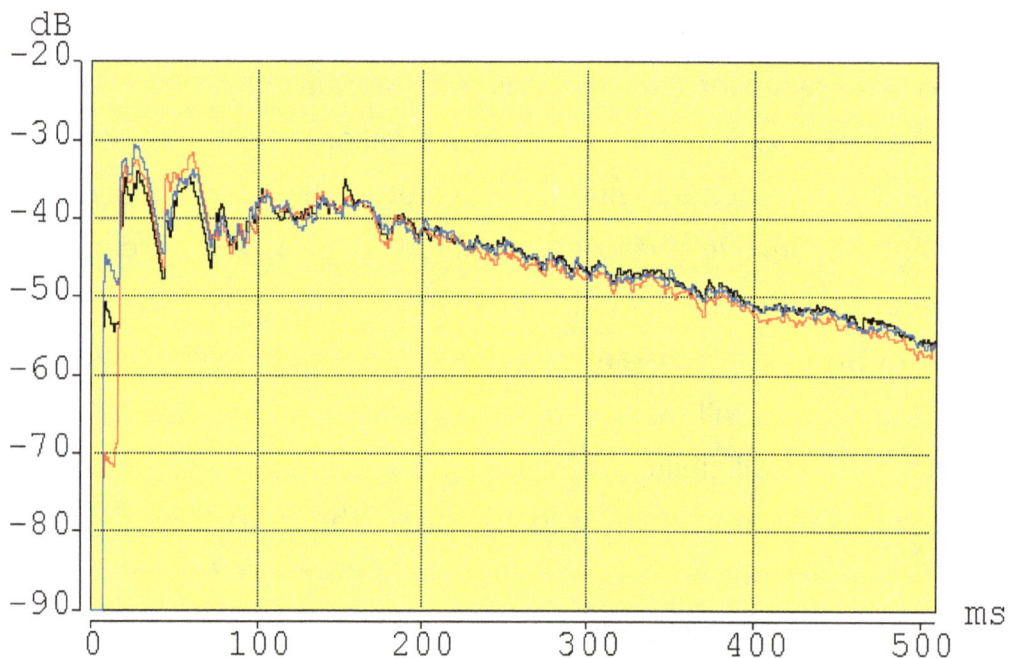

Figure 7-1 -- Sound Decay Curve Showing Early Reflections

8.   Other Factors

Although reverberation time and strength are physical characteristics of a room, other important characteristics including intimacy, envelopment, warmth, spaciousness, and clarity are perceptual; characteristics whose direct measurement involves human listeners. However, research done during the last several decades has done much to establish connections between these perceptual characteristics and the physical design variables that must be manipulated in order to achieve target values of each characteristic. As a result, a competent acoustician can design spaces for speech or musical performance that provide optimal listening experiences. In some cases, however, the owner may desire a facility that will serve for such a wide range of functions that not all can be optimized in a single structure, or at least within budgetary constraints. Three other characteristics of a room that require careful attention are the ability of musicians to hear themselves and each other ("stage support"), ambient noise from exterior and interior sources, and speech intelligibility.

Intimacy

Intimacy refers to the sensation that the talker, singer, or instrument(s) is located physically close to the listener. Intimacy is important for speech and chamber music. Intimacy depends upon the **initial time delay gap**, the time that elapses between the original sound and the first reflection. Thus an audience hearing a singer standing in the middle of a large stage with a high ceiling and no nearby reflective surfaces will not perceive that singing as intimate; whereas, if the singer is on a small, curtainless stage with a relatively low reflective ceiling, the singing will sound intimate. Some researchers believe that the predominance of television as an entertainment medium has brought about an increased demand for intimacy on the part of music listeners, since television programs normally incorporate little of the reflected and reverberant sound of the performance hall, so the early reflections on which the hearer bases judgments of intimacy are those of the very nearby walls, floor, and ceiling of his own living room.

Envelopment

Envelopment is the sense of being surrounded by the music, and is important for almost any type of musical performance. It depends mainly upon the presence of diffuse reflections from the side and rear of the audience. Diffuse reflections are spread out in time, unlike the coherent reflections that are perceived as echoes. Rectangular, "shoebox-shaped" halls like the Boston Symphony Hall usually provide a great sense of envelopment; while fan-shaped halls provide mainly reflections from the front, resulting in less envelopment.

In addition to the need for side and rear diffuse reflections, envelopment requires that the sound striking the left and right ears be dissimilar. In an extreme case, if a listener were located along the center line of an empty rectangular room hearing a singer also located on the center line, but toward the front of the room, the listener would hear side and rear reflections, but with essentially no difference between the left-ear and the right-ear sound, (s)he would not feel enveloped by the music. A measure of the dissimilarity of left-ear and right-ear sound is called the **inter-aural cross correlation (IACC)**. Lower values of IACC (less similarity between left and right) give improved envelopment and are generally preferred. The direction from which reflections arrive at the listener's ears also affects envelopment, with a preference toward about 55° ±20° from the front-to-rear line of the room.

Envelopment is most important for organ, symphonic and choral music; less so for piano, small solo instrument and solo voice; probably detrimental for speech.

Warmth

Performance spaces are sometimes described as "brittle", "thin", "full", or "warm". These terms describe the timbre of the room, and technically are considered as descriptions of the property called **warmth**. In small rooms such as recording studios, the warmth is controlled by the frequencies at which the dimensions of the room produce resonances (**modes**) that reinforce or cancel the sound at particular frequencies. In most worship and performance spaces, these modes occur at subaudible frequencies, and the warmth depends primarily upon the way in which RT60 varies with frequency. This, in turn, depends upon the way in which the acoustical absorption of the room surfaces varies with frequency. Generally, since only very thick absorbers provide much low-frequency absorption, most rooms have an RT60 that increases at lower frequencies. Thus warmth is usually easy to attain. Typically, for most acoustical music, the RT60 in the octave containing middle "C" (256 Hz) can be about 10-15% longer than the mid-frequency RT60 (an octave above middle "C"), and the RT60 of the octave below middle "C" can be about 25% longer than the mid-frequency value. More variation than this can make the room sound too warm or boomy. Less variation can make it sound thin.

As with most acoustic parameters, the importance and degree of warmth depends upon the purpose of the listening space. Ideal speech conditions involve the RT60 being substantially constant across the range of audible frequencies. The same is true for electronically amplified music. At the other extreme, organ and choral music benefit from acoustical warmth.

Spaciousness

Spaciousness is the perception that the sound source is widely distributed across the front of the room, as opposed to a sense of constricted width. This characteristic is sometimes called the "apparent source width" or "sound stage width" in magazine articles. Spaciousness depends upon the prevalence of sidewall (lateral) reflections that are heard within the first 80 milliseconds after the original sound. Thus the factors that contribute to spaciousness also promote envelopment. Spaciousness is primarily important for organ, choral, and orchestral performances. A wide-sounding orchestra (very spacious room) sounds good; whereas a wide vocal soloist can sound abnormal.

Clarity and Definition

Clarity and definition are closely related, in that both are quantified by the ratio of early sound reflections to total sound. Clarity refers specifically to reflections from all directions arriving within 80 milliseconds of the original sound, and is used in assessing music halls. Chamber music especially benefits from proper clarity, as does any sort of percussive music. Excessive clarity can cause a room to sound dry even if it has enough reverberation, since it is the early reverberation to which the ear is most sensitive.

Definition refers to reflections coming from all directions, but arriving within 50 milliseconds of the original sound. Definition is most often used in assessing listening spaces for speech, and in some cases is used as a criterion for speech intelligibility.

Ensemble, Stage Support

Any time a group of musicians and/or vocalists performs together, an acoustical synergy should occur that permits the individuals to feel and sound like an ensemble – a single entity. Achievement of ensemble is a complex acoustical, psychological, and musical phenomenon, but one essential is that the individual performers must be able to hear themselves and each other well and in proper balance. The acoustical piece of the puzzle involves reflective surfaces that bounce the sound of the individual back to him or her, while also bouncing the sound of the other performers back to each individual as well. These beneficial stage reflections must come from surfaces close enough to the performers that the reflection arrive within the Haas interval of 35-50 milliseconds, because otherwise the sound is perceived as an echo which can confuse the timing and synchronism of the performance. Bands utilizing electronic amplification often use monitor speakers to substitute for stage support, and this can work well for groups with just a few members. However, providing stage support electronically for large groups is almost technologically impossible, so acoustical means must be used. This is another function provided by a well-designed orchestral or choral shell. Some spaces utilize reflective acoustical clouds above the stage to provide stage support for the musicians and vocalists.

Ambient Noise

Noise in any acoustical environment can be distracting, can mask pianissimo music, and can decrease speech intelligibility. A good worship or performance space should have background noise levels at or below the sound level of rustling leaves on a calm day (about 25-30 dBA). Ambient noise comes from one or more of four sources: other people in the room, Noise transmitted into the room from other rooms in the building, noise entering from outside the building, and noise generated by the heating, air conditioning, and ventilating (HVAC) system. The first of these noise sources is not entirely within the realm of control, but it should not be either ignored or overestimated in its importance. Rooms with uncarpeted floors are likely to have greater noise from normal foot shuffling of the occupants than are rooms having carpet on the floors. Rooms using movable seating are susceptible to noise from chairs sliding or creaking as the occupants wiggle around. But in a fine concert hall, the audience spontaneously hushes, even to the point of holding their breath, during pianissimo passages of the music.

Noise entering the room from other rooms or from outside the building is often the most distracting type, because it draws attention from the program to the noise source. This type of noise is best avoided by proper design and construction of the building, as it requires the use of massive materials that are difficult or expensive to retrofit.

HVAC noise is typified by hums, roars, air-rush noise, or sometimes buzzes, and are usually fairly continuous. The primary effect of HVAC noise is to mask the desired sound and reduce speech intelligibility. High HVAC noise also causes increased listening fatigue, as hearers have to expend more mental effort to sort the music or speech from the noise. And HVAC noise, in common with other continuous noise, makes hearing especially difficult for hearing-challenged people.

While some causes of excessive HVAC noise can be eliminated during a renovation, other causes are difficult to address except in the construction of a building.

Uniform Sound Coverage

In a perfect listening space, the sound would be essentially uniform in both loudness and quality throughout the space. There would be no "hot spots" or "dead spots". Acoustically, this would mean that nowhere in the room would there be a concave reflecting surface whose focal point would concentrate sound at any place within the room. It would also mean that the walls had sufficient irregularities that reflections would be diffuse rather than creating recognizable echoes. In classical concert halls, the surface irregularities were often created by the placement of statuary within niches in the walls.

In rooms using electronic amplification, uniform sound coverage requires careful design, selection, location, and aiming of the loudspeakers. Since these are nontrivial tasks, uniform coverage from sound reinforcement systems may be as much the exception as the rule.

Speech Intelligibility

The importance of speech intelligibility has been recognized since the time the theaters of the ancient Greeks were built. But not until the advent of the telephone was there sufficient commercial motivation to improve speech intelligibility that serious research was done on the subject. As most school children know, speech is made up of vowel sounds and consonant sounds. If I am speaking and you can perfectly understand all my vowels but none of my consonants, chances are that you will have no idea what I am saying. But if you understand the consonant sounds and could not identify the vowels, you would have a good chance of recognizing at least some of my words. Thus, the first test devised to quantify speech intelligibility focused on recognition of consonants. A talker would pronounce similar syllables from a standardized list, such as "pa, ta, da", and the listeners would write down what they heard. A perfect score would indicate 100% recognition of the consonants. Those consonants not correctly recognized were considered "lost"; thus, the speech intelligibility was rated in terms of %articulation loss of consonants (%ALcons). There are now electronic instruments that can measure %ALcons, and computer software that can predict it, based upon a computer model of the listening space.

The other common metric for speech intelligibility is the Speech Transmission Index (STI). This was derived through an involved process that analyzed the path from talker's mouth to listener's ear as a communication channel, and quantified the imperfections in this transmission path.

Figure 8-1 illustrates the subjective responses of listeners to different values of %ALcons, STI, and definition. Note that neither the ratings nor the correlation

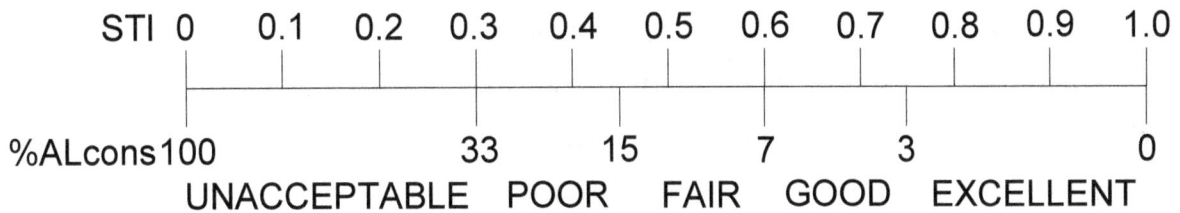

STI 0   0.1   0.2   0.3   0.4   0.5   0.6   0.7   0.8   0.9   1.0

%ALcons 100   33   15   7   3   0
          UNACCEPTABLE   POOR   FAIR   GOOD   EXCELLENT

Figure 8-1: Speech Intelligibility Ranges

between STI and %ALcons is universally agreed-upon. Since the methods of calculating the two metrics are very different, there is no rigorous means of converting STI to %ALcons or *vice-versa*.

Speech intelligibility is affected by the variation of sound level *vs.* frequency, electronically-produced distortion, background noise, and reverberation. If for some acoustical or electro acoustical (sound system) reason, not all frequencies of sound are equally present at the listener's ear, speech intelligibility can suffer. The most critical frequency range is from 2,000 to 4,000 Hz (3 to 4 octaves above middle "C"), since most of the sound energy of consonants falls within these octaves. Distortion (more accurately, nonlinear distortion) is the raspy, buzzing sort of sound caused by damaged speakers, overdriven amplifiers, and electric guitars using "fuzz" devices. This particular cause of poor speech intelligibility was once epidemic in public address systems of bus and train stations. Noise tends to mask the critical consonant sounds, as does excess reverberation, and both of these sources decrease speech intelligibility more severely for hearing-challenged people.

9. Factors in Room Design

Often, when the word "acoustics" is mentioned in connection with room design, all that comes to mind is fuzzy material on the walls. In fact, there are at least a half-dozen factors in the design of a room that affect the listening experience of the presenters, performers, occupants, and audience. As different rooms have different purposes, so the balance of these factors must be carefully controlled during the design process in order to secure a satisfactory final result. For existing rooms that are being renovated in order to improve the acoustical performance, or perhaps to be devoted to a different function that requires a change in acoustics, some of these factors will necessarily be easier to change than others, but proper attention to acoustical factors in design of the renovation will nevertheless yield ultimate benefit.

Shape

Often the shape of a room is chosen with reference only to convenience in planning the building. This is not necessarily bad. Such an approach could, for example, lead to a concert hall of the classic "shoebox" design, such as the Boston Symphony Hall, which is universally acclaimed as one of the best concert halls in the world. However, this hall's design was not an accident of architecture; instead, the hall was deliberately designed after what was perhaps the premiere concert hall in Europe, the Gewandhaus in Leipzig (subsequently destroyed during World War II), and the acoustical design was done by Wallace Clement Sabine, often called the father of modern architectural acoustics. To avoid trapping sound, the balconies were made unusually narrow. The stage walls were banked inward to assist sound projection into the seating area. The ceiling is coffered, providing diffuse sound that prevents acoustical hot spots and dead spots, so that the sound is reasonably uniform from seat to seat.

A room in which listening is an important consideration must be designed so that the shape enhances the listening experience. Figure 9-1 shows a long section of a room as a counter-example. In this church fellowship hall, there is a relatively tall stage house over the stage, and velour theatrical curtains line the top, back, and side walls of the stage and stage house. The audience area is large and rectangular, with an acoustically absorptive ceiling. When someone speaks or sings from the stage, the sound seems to encounter an invisible barrier at the front of the stage, so that the audience cannot hear well. If the talker speaks more loudly, an echo reflects from the back wall and proves very distracting to the person talking. Additionally, a performer or talker on the stage cannot get a sense of the "house sound" because of being acoustically isolated from the audience area. Thus we see a room in which the shape causes the room to act as two separate acoustical spaces (the stage and the audience area), thus inhibiting sound projection; and causes deleterious echoes.

If the stage had been provided with a reflective ceiling appropriately shaped to

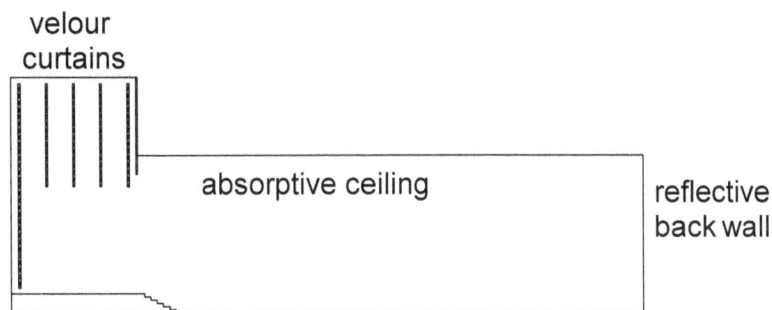

Figure 9-1: Problematic Fellowship Hall

reflect sound into the audience area, the projection of sound from the stage would have been much improved. And if the back wall had been shaped to provide diffusion rather than specular (like light) reflection, the echo problem could have been ameliorated. (Other methods of dealing with the echo will be discussed later in this chapter, in the section on wall finishes.) Figure 9-2 illustrates an improvement to the design that would have helped the acoustics.

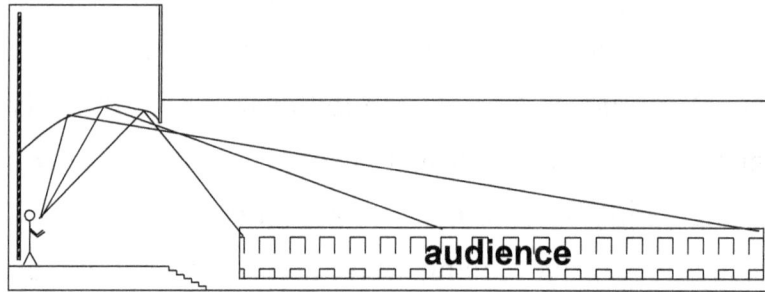

Figure 9-2: Acoustical Projection Improved by Stage Ceiling Shape

In addition to affecting sound projection, the ability of performers and talkers to hear how they sound in the house, and the detrimental effect of echoes, the shape of a room can cause the buildup of sound in specific frequency ranges in certain parts of the room, so that from certain seats, the sound may be harsh, boomy, or otherwise unnatural, and speech intelligibility may be lacking. Figure 9-3 shows the effects of a barrel ceiling in focusing sound on certain spots, making sound coverage in the room uneven. Figure 9-4 shows a similar focusing effect caused by a domed ceiling. Figure 9-5 shows an auditorium with careful acoustical design that gives a helpful reflection of sound to all parts of the auditorium. This same sort of effect is often achieved by the addition of acoustically reflecting "clouds" suspended from the ceiling to provide reflections in the desired directions.

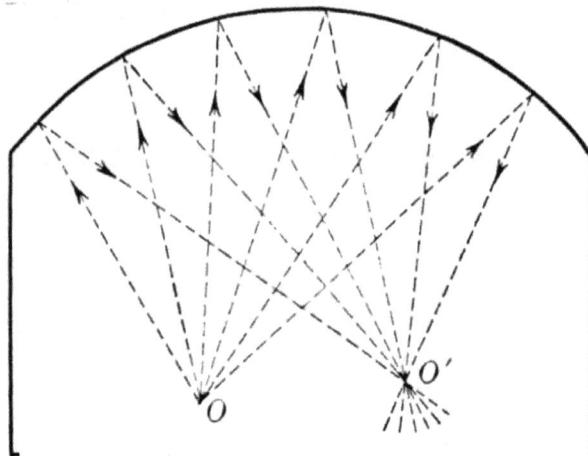

Figure 9-3: Focusing Effect of Barrel Ceiling (After Knudsen)

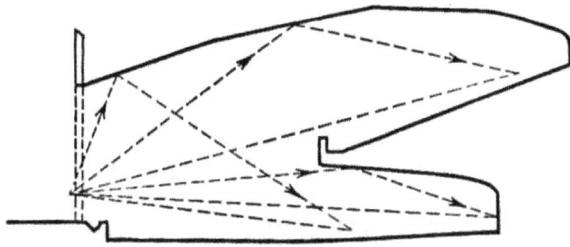

Figure 9-4: Focusing Effect of Domed Ceiling (After Knudsen)

In a rectangular room with a raised stage and little or no rake (slope of the seating area), echo from the back wall is a common problem. If the length of the room is greater than about 25 feet, reflections from the back wall can be perceived as echo, and the longer the room is, the more annoying will the echo be. Room shape can ameliorate echo in either of two ways. If the seating is raked, the sound traveling across the leads of the listeners will strike the back wall at such an angle that the reflection will travel well above the listeners' heads. This will change the echo problem, although there may still be troublesome secondary reflections from the ceiling and front wall. If the back wall is irregularly shaped, the reflected sound will be diffused so as not to produce a coherent echo. A concave rear wall is not an acceptable shape: not only will it not eliminate the echo, but it can cause focusing problems as illustrated earlier in connection with domed ceilings.

53

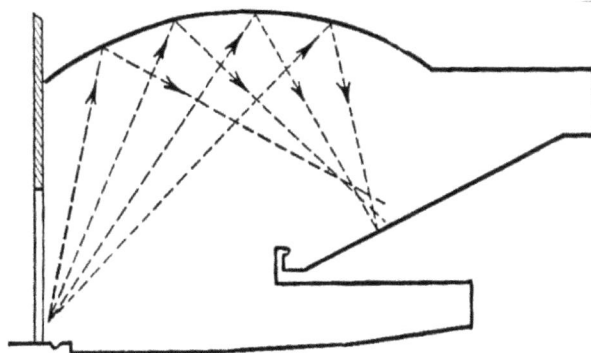

Figure 9- 5: Even Sound Coverage from Proper Ceiling Contour (After Knudsen)

Another type of echo that can be problematic is so-called "flutter echo" caused by sound repeatedly reflecting between parallel walls. Figure 9-6 shows a wide stage with parallel walls and little sound absorption between them. The graph below the drawing illustrates the flutter echo produced by a handclap at the position of the small circle. The sound impulse is repeated at

56-millisecond intervals until it eventually dies out. In this case, flutter echo could be prevented if the walls were not parallel.

Flutter echo can be observed in many empty auditoriums, if one claps the hands while standing in the center of the seating area. However, in this case, it is not usually a problem, since the main sound source is not usually located between the walls, but rather at the front of the auditorium. Furthermore, the presence of the audience normally provides acoustical absorption that rapidly attenuates any echoes. The problem illustrated in Figure 9-6 occurs because the sound source is between the parallel walls, and almost no absorption is present. A particularly bad situation would be one in which a percussion set was placed between parallel reflecting walls.

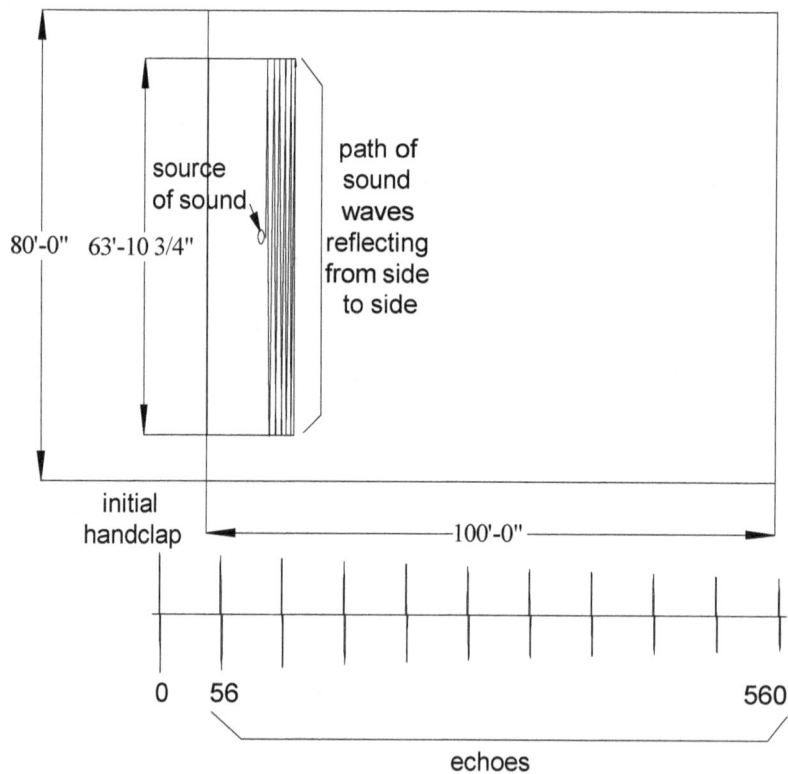

Figure 9-6: Flutter Echo

The last acoustical factor related to shape is one that is not important in worship and performance spaces, but is mentioned because many people are somewhat acquainted with the phenomenon. The well-known "singing in the shower" effect caused by room resonances results from what are called standing waves that occur at specific "modal frequencies". When the maximum-pressure point of a sound wave occurs at the same point on each trip as it is reflected around a room, the sound at that frequency will build up, causing a "hot spot". If the maximum-pressure point aligns with the minimum-pressure point of a preceding wave, sound cancellation will occur, producing a "dead spot". In any room, the modal frequencies at which hot spots and dead spots occur will be spaced far apart at low frequencies, becoming closer together at higher frequencies, until above the "Schroeder frequency", they merge into a continuous band, and have virtually no effect. For example, a typical living room may have its lowest modal frequency at 45 Hz, and within the octave from 45 to 90 Hz there may be a total of eight modal frequencies. In the 180-360-Hz octave containing middle C, there may be 25 modal frequencies. In this small room, the Schroeder frequency is perhaps 220 Hz. In essentially any worship or performing space, the Schroeder frequency will be low enough that room resonance is no problem even at the lowest audible frequencies.

Volume

In addition to numerous architectural considerations affecting the choice of enclosed volume (cubic feet or meters) of a worship or performance space, there are three important acoustical factors that should be considered. The first is the needed reverberation to support the type of music or style of worship to be employed in the room. As discussed in Chapter 5, the reverberation of a room depends upon the area of all surfaces in the room, the total acoustical absorption of the room, and the enclosed volume of the room. Thus a room in which organ and choral music are to be performed will benefit from a fairly long reverberation time. A long reverberation time requires a large volume. Since the volume equals floor area multiplied by height, and floor area is often determined by the desired seating capacity, the needed volume translates directly into a minimum ceiling height. For example, a room having internal surfaces totaling 88,000 square feet, with an average acoustical absorption of about 28%, will require a volume of 1,200,000 cubic feet in order to provide a reverberation time of 2 seconds. If the floor area is 150' X 200', the required height will be 40'. In an actual room design, decisions about volume are also affected by the presence or absence of balconies, as well as room shape, since these will impact the total surface area inside the room, as will the height: there are multiple interacting variables to be considered.

A second acoustical factor related to total volume is the strength (discussed in Chapter 7) or "room gain", or ability of the room to support unamplified speech and music. Actually, it is more specifically the room height that affects room gain. When a sound is produced at the front of a room, some of the sound travels directly to each listener, and some is reflected off various surfaces in the room. It is the effectiveness of these surfaces as reflectors that constitutes room gain. Sound that reflects off the side walls is attenuated by its passage across the heads of the audience, but sound reflected off the ceiling can be important in reinforcing the desired sound level. If the ceiling is very low, it is unlikely to provide useful reflections to the back of the audience. If it is too high, the reflections may be sufficiently delayed behind the direct sound as to constitute an echo. But there is a range of optimum ceiling height from the standpoint of room gain for a listening room of any floor area and shape.

The other acoustic factor related to room volume also depends more directly upon height, which of course affects volume. Since most worship and performance spaces are provided with sound systems, it is wise to consider the requirements of the sound system during the planning of a room. Rooms with very low ceilings (8 – 12 feet) can be served by flush-mounted ceiling speakers if speech and vocal music are the main elements to be amplified, and the needed sound levels are not over about 70 – 80 dBSPL. For situations involving higher sound levels or music having high bass content, other types of speakers will be needed, and these will be difficult to arrange so as to provide even coverage of sound throughout the space. In addition, such speakers will be highly visible, which may create an aesthetic consideration for some owners. From the perspective of sound system design, 20′ is a desirable minimum height for a small-to-medium room (seating capacity up to about 300).

Occupancy

In most worship and performance spaces, the congregation or audience comprise the primary acoustical absorbers. Thus the reverberation time of the room will depend upon the number of people in the room. Any musician will attest that rehearsing in an empty auditorium is challenging because of the extreme reverberation. Readjusting one's performance for the acoustically deader effect of an occupied hall requires skill. Some facilities, such as concert halls in large cities, are expected to operate at or near full occupancy most of the time. Other facilities, such as some church sanctuaries, may operate at one-third or less occupancy for a substantial number of events, being fully occupied only two or three times per year. It is essential that the expected occupancy be taken into account during the acoustical design of the room, so that the results will be satisfactory for most of the room's uses.

Large meeting spaces, such as church fellowship halls, are often used for both meal service and lectures, concerts, or drama, within the same event. In this case, greater occupancy translates directly into greater ambient noise, since the more people there are eating, the greater the noise that will be produced. Minimizing this ambient noise requires careful acoustical design, and the level of the ambient noise is a factor to be considered in sound system design. Again, good acoustical design involves accurate prediction of the average occupancy of the room during use.

Finish Materials

As discussed in previous chapters, there are two ways in which sound can be reflected from a barrier: specularly, and diffusely. Specular reflection is like the reflection from a mirror: specular reflection of light produces a mirror image, and specular reflection of sound produces an echo. Diffuse reflection retains most of the intensity of the incident wave, but without the details. Diffuse reflection of light produces bright spots of the same color as the incident light, but no image; and diffuse reflection of sound retains the pitch and some of the intensity, but the timing of transients is smeared. Thus diffusion does not decrease reverberation, but it can prevent the formation of echoes. Just as specular reflection of light requires the smooth reflecting surface of a mirror, so specular reflection of sound requires a relatively smooth surface. More specifically, any irregularities in the

Figure 9-7a: RPG QRD Diffusor for Walls (Courtesy RPG, Inc.)

Figure 9-7b: RPG Golden Pyramid Ceiling

Diffusor (Courtesy RPG, Inc.)

surface must be small with respect to a wavelength. It follows that to achieve diffuse reflection of sound, a substantially irregularly-contoured surface is required. The older concert halls and opera houses of Europe used niches in the side walls, housing statuary. In addition to providing a visual design element, these also created diffuse reflections.

Rear walls and balcony faces in worship and performance spaces often need to provide sound diffusion in order to prevent echoes. Over the years, many approaches to wall design have been used with greater or lesser degrees of success, including building the wall into a convex shape, employing smaller convexities within the wall, and creating geometric designs such as coffers in the wall surface. During the last several decades, diffusive wall panels have been introduced by a number of manufacturers. (See Figure 9-7 for two representative examples out of many different available products.) These can be used to provide a wall or ceiling surface with carefully engineered diffusion.

Aside from the ability of walls to provide acoustical diffusion, the acoustical absorption of walls is very important. All materials provide acoustical absorption, and the absorption is always dependent upon frequency. Most materials provide increasing absorption as frequency increases. Soft materials, such as carpets and draperies, often come first to mind in connection with acoustical absorption. However, both of these absorb mainly high frequencies, having little effect at or below the male voice range. Even brick, stone, plaster, and concrete walls provide some small absorption at high frequencies. Gypsum board provides some high-frequency absorption and also absorbs sound in the frequency range about an octave below middle C. This occurs because of the panel flexing and using up

59

acoustical energy in the process. The exact amount and frequency of this "diaphragmatic" absorption depend upon the thickness of the gypsum board, the depth of the cavity behind it, and the presence or absence of absorptive material in the cavity between the studs. Thin wood paneling provides significant absorption at low frequencies, by the same diaphragmatic process. There is a material called "acoustical plaster" that provides much greater absorption than ordinary plaster, and that has been used with success in many rooms. A good acoustical consultant will know of many other acoustical materials, as new products are being introduced all the time. Figure 9-8 shows a typical fabric-wrapped fiberglass wall panel designed to provide acoustical absorption. Where aesthetic considerations dictate avoiding the appearance of separate panels, s very similar product is available "built in place". In this process, the fiberglass is attached to the underlying wall surface, then a seamless piece of fabric is stretched over it, giving the appearance of a fabric wall covering. "High-Impact" fiberglass wall panels having a surface layer of thin rigid fiberglass board are available for areas where the panels may be subject to damage. These high-impact panels are typically sized as 1 1/8", 2 1/8", etc., rather than in whole inches, as the rigid board adds about 1/8" of thickness. These products provide better low-frequency absorption than do standard fiberglass panels.

Figure 9-8: Acoustical Solutions' AlphaSorb Fabric-Wrapped Fiberglass Wall Panel (Courtesy Acoustical Solutions, Inc.)

60

A less expensive material called Tectum™ can be used for many applications, especially when excellent low-frequency absorption is not essential. This material is available as bare panels, having the look of tangled wood fibers (which is what they are), or fabric-wrapped for applications where appearance justifies more expense. The absorption of Tectum can be strongly enhanced by adding fiberglass or other fibrous absorber in a cavity behind it, and deeper absorptively lined cavities can increase the low-frequency absorption.

Suspended ceilings require careful analysis in their effects on acoustics. Somehow, lay-in ceiling tiles have acquired the generic name "acoustical tile", even though many of these products do not provide much acoustical absorption. Reference to the technical literature provided by the tile manufacturers will reveal a wide range of acoustical properties among which one can choose. It is also important to remember that the details of mounting the suspended ceiling, and the presence or absence of absorbent material between the suspended ceiling and the rigid ceiling above may greatly affect the performance. If the ceiling panels are porous such as fiberglass or Tectum, then adding absorber above will enhance absorption. A major benefit is obtained for Tectum. Little if any benefit is seen with mineral fiber ceilings. Ceilings are often the least expensive place to add acoustical absorption, and there is little visual difference between an absorptive lay-in ceiling panel and one that is not absorptive. However, since reflections from the ceiling are important for good speech intelligibility, intimacy, and clarity, the advice of a good consultant should be used in any decisions involving adding absorption on a ceiling. Usually, in non-performance rooms such as cafeterias, lobbies, and libraries, absorptive ceilings are appropriate.

Floor finishes have two acoustical effects: noise production and acoustical absorption. Hard floors in meeting rooms can cause unnecessary amounts of noise due to people walking around. In auditoria, foot-scuffling noise can be distracting during quiet musical passages. Carpet alleviates both of these issues, and also reduces reverberation, which may or may not be desirable in a given room. Different carpets provide differing acoustical effects. Thick carpeting, or carpet installed over an underpad, provides greater acoustical absorption, especially for frequencies at and below the male vocal range. Several companies make "acoustical" cork or cushioned vinyl floors that can help to reduce foot noise.

However, these floors should not be expected to provide acoustical absorption as carpet would do.

Although as previously mentioned, draperies absorb primarily high frequencies, they should be considered in any acoustical analysis. Heavy velour stage curtains in particular can provide a significant reduction of the helpful reflections from stage walls; or if used on other room surfaces, they can help somewhat with reverberation control. Lighter draperies such as those commonly used as window treatments, or thin decorative banners, have little acoustical effect. It is also important to remember that not only the absorptivity of a material, but its surface area is important in the overall acoustical effect it provides. A square yard of 100% absorptive material will have less effect upon diminishing reverberation than will a 10' X 30' brick wall section.

Acoustical Insulation

The complete acoustical design of any space consists of two parts: control of unwanted noise, and provision of desired acoustical characteristics for the program material in the space. The former part – control of unwanted noise – can also be split into two parts: prevention of noise intrusion from outside the space, and prevention of excessive internally generated noise. The prevention of noise intrusion from outside involves choice of acoustic insulation. In spite of the common name, acoustical insulation and thermal insulation have only function in common. Both serve to keep a certain kind of energy within, or outside of, desired areas. It is true that thermal insulation such as glass and mineral fibers provide acoustical absorption, but these materials are not in fact acoustical insulators. Acoustical absorption in the cavity of a partition can increase the sound insulation of the partition. Acoustical insulation consists of massive materials such as concrete walls.

Acoustical insulation is specified in two ways: for walls, it is specified as "sound transmission class" or STC, and for floor-ceiling systems, it is specified as "impact isolation class, or IIC. Both are specified as numbers, with a higher number indicating greater acoustical insulation. Typical single-family residential construction provides walls with an STC of 30-40, and typical floor-ceiling systems provide an IIC in the 25-35 range if the floor is hard. Higher values of STC and IIC are designed into partitions and floor-ceilings between units of multifamily structures: a minimum of 50, and up to 60 in wood frame structures, or higher in concrete construction. Situations that may indicate a need for greater acoustical insulation include very noisy outdoor exposures such as jet aircraft, major highways, or outdoor sports or performance venues, or noisy adjacent indoor rooms, such as gymnasia, bars, or mechanical equipment rooms. In addition to the wall structure, doors and windows need to be carefully analyzed for acoustical insulation. Indeed, there have been situations in which HVAC equipment located just outside a church sanctuary provided unacceptable noise intrusion through a stained-glass window, even though the walls had a high enough STC. The primary contributor to STC is mass, although multilayer partitions whose layers are resiliently isolated from each other can produce substantial improvements in STC as well.

Noise isolation using acoustical insulation has often been called "soundproofing", but most experts in the field avoid that terminology. The reason is similar to the philosophical concept called "Zeno's paradox", which points out that one can never reach a destination by repeatedly moving half the remaining distance. We can reduce the noise entering a room by some percentage, but not by 100%.

Rooms requiring careful attention to IIC not include those in which activities in an adjacent room above the space can produce noise from footfalls or moving furniture. An example is classrooms in which chairs and tables being moved in an upstairs room must be prevented from interrupting classes in a downstairs room. The factors determining IIC are the same as those that determine STC, with the addition of the resiliency of the floor material. Hard shoe heels impacting a cork floor will produce less noise in the room below, than would the same heel impacts on marble tile. Thus the cork flooring will increase IIC for the floor-ceiling system. Carpet strongly increases IIC. For hard surface floors, a resilient material is required under the floor surface if an high IIC is to be obtained, isolating the hard floor from the subfloor.

The main factor in providing a high STC is mass (or weight), or specifically mass per unit area. A concrete wall having a surface weight of 39 pounds per square foot will provide an STC of 47; whereas, a 2"X2" wood-framed wall faced on both sides with ½" gypsum board has a much smaller surface weight of 5.9 pounds per square foot, and a lower STC of 32-35 . Fiberglass batts placed in the cavity between wall faces will increase STC slightly, because of sound being transmitted through the studs. However, the STC can be increased substantially by both isolating the gypsum from the studs on one side and putting acoustical absorption in the cavity. For commercial construction, 25 gauge studs adequately isolate the gypsum, but heavier studs must be treated as described above for wood studs. Products are available to isolate the gypsum from the studs. For very high performance, separate studs can be used for each side.

Good sound isolation requires good air seals. Walls that extend from a concrete floor to a corrugated roof deck often have so much sound leakage at the top that no attempt at improving noise isolation will do much good unless the gaps are sealed with acoustical caulk.

There are two ways in which noise isolation efforts often fail. One is shared HVAC ducts. In one specific case, a recording studio was built into an existing building, and great care was used in the design of walls, floors, and ceilings. Yet when the studio opened for business, the owner found to his chagrin that noise from adjacent offices traveled through the HVAC ducts into the studio, nullifying his efforts at eliminating noise intrusion. Separate HVAC systems should be used for noise-sensitive spaces.

The other pitfall for noise isolation efforts is called "flanking paths". A flanking path is any way in which sound can travel without passing through the partitions that have been designed to block noise. One example is concrete slab floors through which noise generated in one room can be conducted and re-radiated in an adjacent room. This most often happens with impact sound rather than airborne sound on concrete. The problem for airborne sound is much worse when a wood floor surface is continuous between adjacent spaces, or a metal deck is continuous, or even a piece of gypsum on a wall is continuous. The solution to problems involving conductive flanking paths is resilient barriers isolating the floor on one space from that of adjacent spaces. The main solution is to not allow a surface to extend from one space to another without a break, or if that is not possible, then to cover that surface with another which does not extend from space to space.

65

A second example of flanking paths is suspended ceilings combined with walls that do not extend to the roof deck. In this case, noise travels through a fairly low-STC ceiling out of one room to the shared area over the ceilings, and then down through a fairly low-STC ceiling into other rooms.

There are many materials that are advertised to provide superior "soundproofing" or noise isolation, but each of them will work properly only in certain kinds of applications, and then only if installed in the proper fashion.

Heating, Ventilating and Air Conditioning (HVAC)

The major contributor to interiorly generated noise in most rooms is the HVAC system. A competent mechanical engineer can design an HVAC system that is literally so quiet that one has to hold one's breath to hear it in operation. The noise level in a room is specified by a number called the "noise criteria", or NC. A THX™

movie theater is designed to a maximum noise of NC20; a typical theater, NC20-25; deluxe private residences, NC20-30; upscale restaurants, NC40-45; and concert and recital halls and recording studios, NC15-20.

Quiet HVAC design requires large distribution and return vents so that low air velocity can be used. It also involves lined ducts or silencers to attenuate fan noise, and sometimes acoustically insulated ducts to prevent noise within the ducts from "breaking out" into the worship or performance space. Where multiple rooms are connected by ducts, careful attention to duct design is necessary in order to prevent noise from one room from intruding into other rooms. Quiet HVAC design should be incorporated from the inception of a project, as retrofitting existing systems to reduce HVAC noise is very difficult and sometimes impossible.

Factors Affecting Hymn Singing

A very important acoustical concern for worship spaces is the extent to which the acoustics of the space support congregational participation in hymn singing. Part of this support comes by way of the space having the appropriate amount of absorption. However, a factor at least equally important is the ability of parishioners to hear each other sing. Most people are very uncomfortable if they feel that they are singing alone in a public place. In a room with too much absorption, or with acoustical design that does not reflect the sound of the congregation back to themselves, this illusion can prevail. The physical structures responsible for providing such reflection include adjacent walls and ceilings, uncarpeted floors, and uncushioned pew backs and seats. Where these structures are absent, remote, or insufficiently reflective, congregational hymn singing will be poorly supported. Ceilings in worship centers should generally be reflective in order to provide adequate room gain. Even so, other considerations often require a ceiling height so great that reflections from the ceiling are too weak to support hymn singing very much, though the room might be very reverberant due to the high volume. In this case, other reflecting surfaces must be provided. Traditional rectangular rooms that are significantly longer than they are wide have side walls that can provide beneficial reflections. Sanctuaries are often designed in a fan shape in order to increase the sense of visual intimacy for as many worshippers as possible. However, the fan shape results in side walls being located remotely from the congregants.

In many worship centers, the pews are cushioned for reasons of comfort. Pew cushions also have the advantage of reducing the change in a room's acoustical environment resulting from variations in occupancy. This reduces the variation in acoustical support for the musicians and choir in rehearsal versus in a worship service. However, the pew is the closest reflecting surface to a congregation member, and therefore an important source of beneficial reflections. Studies have shown that pew seat cushions only slightly reduce the ability of congregants to hear each other sing.[Honeycutt] Pew back cushions, however, reduce the beneficial reflections more severely, since the pew back is closer to both the ears of a singing congregant and the mouths of his/her singing neighbors. Some worship centers have been built with carpet in the aisles but not under the pews, in an effort to preserve support for hymn singing. Computer models show that the presence or absence of carpeting under the pews has little effect on reflections that support hymn singing, and also little effect on the reverberation time of the room.

In some worship centers, the need for a high ceiling to provide a long reverberation time can create design conflicts in the effort to support hymn singing. One approach that has been suggested is to build the outer shell of the room with enough volume for a long reverberation time, but to install a low reflective ceiling over the seating area to provide beneficial reflections for congregants. This ceiling is separated from the side walls by a gap of a few feet so that the space above is acoustically coupled to the main space in the sanctuary. In effect, this sub-ceiling amounts to a large acoustical cloud similar to those used in concert halls. This design also has the advantage of providing good clarity for high speech intelligibility, even in the presence of a long reverberation time. For optimum results, the design of the sound system must be carefully coordinated with the architecture when this design approach is used.

Attaining the Goal

10. Preferred Acoustical Performance of Performance Spaces and Sanctuaries
General Comments

As has been pointed out several times in this book, the preferred acoustical performance of any space depends upon the program material to be used. Performance spaces can be categorized as music halls, drama theaters, lecture halls, or multipurpose auditoria. Within the category of music halls, there are opera houses, chamber music halls, symphony halls, and jazz/rock/pop music spaces. Sanctuaries share similar subcategories, depending largely upon the style of worship music. However, sanctuaries also need to perform well for speech; thus some compromise is often needed. The classification of worship music styles has involved some effort for generations. The early acoustical texts divided worship music into Catholic styles (organ and choral music) and Protestant/Synagogue styles (much like general-purpose concert halls). In the last three decades, much diversification has occurred, so that now classifying by denomination is not instructive. Neither is classification as "traditional" or "contemporary", since hymns of the late 19th and early 20th century are considered traditional by some people, while to others, "traditional" means Bach; and to still others, "traditional" means one of the Gospel styles. Even within Gospel styles, there are "Nashville Gospel", using electrical instruments and percussion; "old-time Gospel", using mainly piano and/or guitar, banjo, mandolin, etc.; and African-American Gospel", which probably includes Hammond organ accompaniment and percussion, along with any of a plethora of other electrical, acoustical, and percussion instruments. "Contemporary" can mean the "folk Gospel" styles on the 1960's and later, light-pop styles including many praise choruses, or Christian Rock.

In order to avoid excessive wordiness, we shall adopt category names that correspond to the musical roots of the various styles, as follows:

Medieval – Gregorian chant with pipe organ

Baroque – Bach, Handel, Haydn, and similar musical styles (more and faster motion that Medieval)

Classical – Mozart, Telemann, and similar musical styles (more and faster motion than Baroque) Acoustically, this style includes the requirements for most 19th- and 20th-century hymns.

Folk/Jazz – Music similar in style to British, Irish, Scottish, and American folk music. Acoustically, this style includes "old-time Gospel", "folk Gospel", and light pop sacred styles.

Rock – Electronically amplified music with strong percussive elements. Acoustically, this also includes "African-American Gospel", "Nashville Gospel", and "Latino" style music.

Speech – Rooms used only for speech.

Multipurpose Venues – Rooms used for a broad spectrum of musical styles

Categorizing worship music styles in this manner also facilitates comparison between the acoustical behavior of worship spaces and that of performance spaces.

The above category listing is in order of decreasing RT60, from 5-7 seconds for Medieval to well under a second for Speech. As mentioned earlier in this book, the preferred RT60 also depends upon the room volume, with a longer RT60 needed to make a large room sound right to the listeners.

A requirement for good speech intelligibility in performance and worship spaces is somewhat eased by the ubiquity of sound reinforcement systems. However, since a long reverberation time is detrimental to speech intelligibility, the cost of the sound reinforcement system is also affected by the RT60, especially for values above 1.5 seconds (Classical style). Values over 2 seconds may require heroic measures in order to obtain good intelligibility.

Since the occupancy of most venues varies widely from one use to another, it is very beneficial for the RT60 to remain reasonably constant in spite of these changes. An excellent chamber ensemble performing on a snowy night in a very good room having an RT60 of 1.3 seconds when fully occupied may be charged with a less-than-adequate performance if the scanty audience results in the RT60 increasing to 1.9 seconds. The sanctuary that sounds wonderful on a high holy day (with full occupancy) may be cave-like and exhibit poor clarity and speech intelligibility when almost empty. The stability of RT60 in spite of occupancy changes is also important to musicians who must practice in an empty room and perform in a full one. Since people are acoustically very absorptive, the only way to stabilize the RT60 against occupancy variations is to prevent the people's absorption from dominating the response of the room. In most concert halls. This is accomplished by the use of cushioned seats. A medium- or heavily-upholstered concert chair has nearly the same absorption whether occupied or unoccupied. In a worship space, pew seat cushions can be used, but the use of pew back cushions tends to reduce the helpful reflections needed for strong hymn singing, so most consultants recommend pew seat cushions only, knowing that this compromise between stable RT60 and strong hymn singing is necessary.

The stability of RT60 is less problematic in rooms with shorter RT60, since the listeners do not contribute as high a proportion of the total acoustical absorption in such a room. (Short RT60 means that the room is more absorptive empty than the same-size room with long RT60.)

Performance spaces in which unamplified vocal solos are to be performed, and worship spaces in which unamplified talkers need to be understood, should also display good acoustical strength, requiring excellent early reflections of sound to the listeners. In a completely open outside space, a listener located 16 meters (about 52 feet) from a person talking at conversational level will experience a sound level of about 41 dBSPL – about the level of a quiet room. In a room having an acoustical strength of 10 dB, the listener would experience a level of about 51 dBSPL, which would sound twice as loud. Since strength, clarity, and definition all depend upon the same physical features – surfaces that provide strong early reflections—an improvement in one shows up as improvement in all three. In fact, improving the early reflections also assists speech intelligibility as well. You will recall that provision of good early reflections is largely a function of the shape of the room, especially the stage (chancel in a worship space) and ceiling. This is one reason that planning for good acoustics cannot wait until the architectural work is more-or-less finalized; it needs to be done early in the process.

In order to promote strength, clarity, definition, and speech intelligibility, ceilings of most performance and worship spaces should be acoustically reflective. Acoustical absorption, when necessary, should be added at other places.

Stage support is important in both performance spaces and worship spaces, since musicians and vocalists need to be able to hear themselves and each other no matter what the venue.

Intimacy and warmth are also important in both performance and worship spaces, but envelopment and spaciousness pertain more to a good concert experience than to worship. In fact, too much envelopment is sometimes distracting, especially in the folk/jazz or rock style sanctuary.

It should go without saying that control of ambient noise is essential to a good listening or worship experience.

The careful reader will have noticed that attaining the desired acoustics on a performance or worship space involves balancing many conflicting parameters. Many building owners have been the victims of designs by professionals who failed to appreciate the complexity of the acoustical design problem, and created a room in which, for example, the RT60 was ideal, but the clarity and speech intelligibility were at best inadequate. For this reason, owners should choose their acoustical consultant with at least as much care as they apply to the choice of an architect or builder.

Medieval

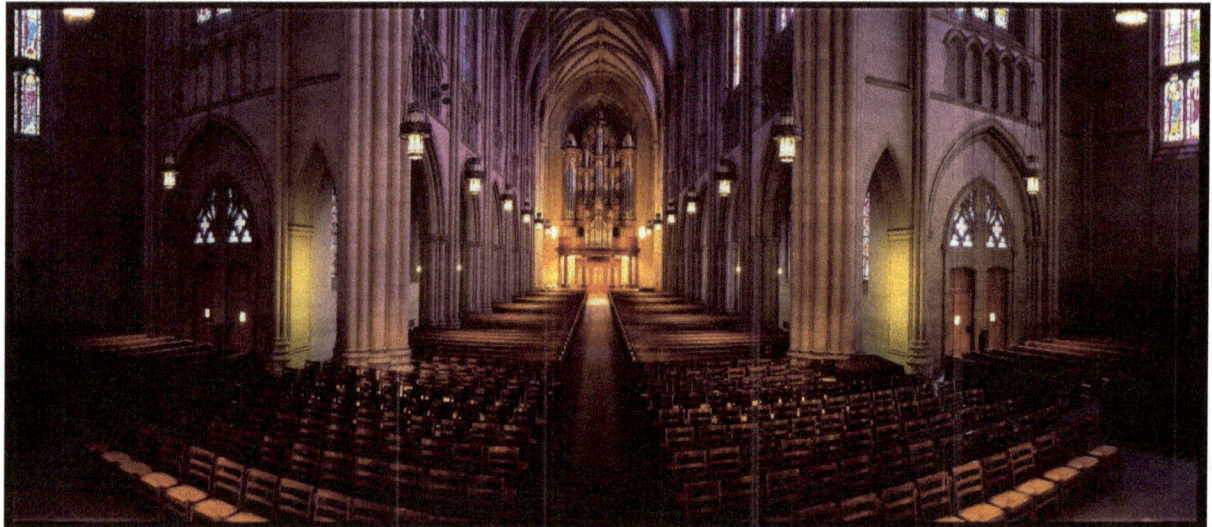

Figure 10-1 Duke University Chapel (Used by permission of Duke University).

Medieval-style venues (sometimes referred to as "Gothic Cathedrals") comprise very large spaces primarily used for performance of pipe organ and choral music. These spaces may have RT60's of 5 to 7 seconds, and as such, they are suitable for slow-moving music such as Gregorian chant. Even pipe organ music of the Classical and Baroque periods, and especially more modern or contemporary organ repertoire containing rapid passages (e.g., Widor's "Toccata") are not well supported by such a long RT60. Speech intelligibility needed for announcements or liturgy typically requires heroic efforts on the part of very competent sound system designer, and may involve measures such as pew-back speakers or a distributed directional system. Depending upon local regulations, fire safety requirements may mandate a sound system having some specific level of speech intelligibility, even if program requirements do not involve much speech.

The high acoustical reflectivity of most surfaces (often stone or concrete) in a Medieval-style venue can result in an RT60 well above the target value when the room is unoccupied. When the room is occupied, audience absorption will bring the reverberation down quite a bit. Reducing the reverb variation with varying degrees of occupancy is most satisfactorily done by the use of seat cushions. In the unoccupied condition, these cushions provide much of the absorption that would otherwise be provided by an audience, and in the occupied condition, the cushions are covered by the audience and add very little to the absorption; thus, there is less variation in reverberation if seat cushions are used. A competent acoustical consultant will always predict RT60 for empty, partial, and full occupancy.

The typical materials used in a Medieval-style space have very low acoustical absorption that varies little with frequency, so that there is enough low-frequency reverberation to provide warmth. Most materials exhibit a slight increase in absorption at high frequencies, and Medieval-style venues are typically large enough that the long sound paths provide significant high-frequency absorption by air. Thus there is usually a slight decrease in RT60 at high frequencies. We are accustomed to the sound of this phenomenon, so it sounds fine to us for music.

The strong reverberant reflections usually mask discrete echoes, so they are seldom a problem in a Medieval-style venue. They also naturally provide great envelopment and spaciousness, although intimacy, clarity, and speech intelligibility are very low, unless electronic amplification is used to enhance these qualities for specific program material. Because of the great acoustical strength, the effects of ambient noise are exacerbated in a Medieval-style venue, so great care needs to be taken in that regard.

Although singers will certainly be able to hear themselves well in such a space, much of what they hear may be "hall sound", lacking the clarity needed for good stage support, so it is very important to provide excellent early-reflecting surfaces in close proximity to the performers.

Baroque

Figure 10-2: First United Methodist Church, Lexington, NC. (Photo used by permission.)

Baroque music is typified by the compositions of J. S. Bach, G. F. Handel, and Antonio Vivaldi. These compositions were affected by the environment in which the composers worked. Bach's church (Thomaskirche in Leipzig), is thought to have had an RT60 about 1.6 seconds. The migration of church architecture toward styles having less reverberation came about largely because of the Protestant emphasis on the spoken word in worship services, as opposed to the ritual chants in a foreign language used in Catholic services. This much shorter reverberation time allowed Baroque music to be more highly figured than medieval music, which was written for Gothic cathedrals. Likewise, Handel's oratorios were often performed in rather small halls (seating about 400-600) having RT60's about 1.5 seconds. Therefore, Baroque music (and Italian opera, as well) sounds best in venues having an RT60 of 1.5-1.6 seconds.

The room dimensions and material choices that lead to an RT60 in this range also lead to good clarity, intimacy, and speech intelligibility if proper attention is paid to creating surfaces for early reflection of sound from performers to the audience. Proper shaping of stage houses can provide good stage support as well. Envelopment and spaciousness require diffuse sidewall reflections which can be provided architecturally (as in older venues) or by use of specifically designed diffusing surfaces. Many performance and worship spaces built since the 19th century make heavy use of gypsum wallboard or wood paneling rather than the plaster that was common in older buildings. These materials can have excessive acoustical absorption at low frequencies, even extending into the bass vocal range. The result of this can be too low RT60 and acoustical strength at those frequencies, making the room sound "thin".

A low RT60 will not mask echoes very well, so Baroque rooms need careful attention to diffusing or absorbing surfaces on the rear and side walls to prevent "slap" and "flutter" echoes. If a Baroque room is built with the correct RT60 and proper attention to clarity, strength and prevention of echoes, even unamplified speech will often be reasonably intelligible, and design of a sound reinforcement system will be straightforward.

Classical

Figure 10-3: First Baptist Church, Lexington, NC. (Photo used by permission.)

Classical music is the category that encompass Mozart, Haydn, Beethoven, and Schubert. This music tends to be less complex than Baroque music, and often consists of a melody over an accompaniment of chords. Many traditional hymn melodies are classical in style, so one definition of a "traditional" worship style is one that involves classical-style music. (Of course, since different people have different traditions, the word "traditional" has various meanings.) In the development of music, the Classical and the later Romantic eras are not well defined, but have been called a continuum of development. Indeed, although Beethoven's early music is considered Classical, some of his later music, especially the sixth and ninth symphonies, are considered Romantic.

Classical and Romantic music encompasses both chamber music, which is usually performed in an intimate setting having an RT60 about 1.4 seconds, and concert repertoire for larger halls, typically in the 1.7 second range for Classical and extending to about 2 seconds for Romantic music. In the Romantic period, some of the fine detail in the music became less common, and wide dynamic range and a passionate feel replaced it; hence the decreased need for clarity (due to shorter RT60) and increased need for the emotional charge resulting from a longer RT60.

The comments on other acoustical properties made in the section on Baroque spaces also applies to Classical/Romantic spaces, with the reminder that when clear speech is important and a longer RT60 (over 1.6 seconds) is chosen, architectural design needs to incorporate careful provisions to maximize clarity and speech intelligibility, even though a good sound reinforcement system is to be employed.

Figure 10-4: Pinetops United Methodist Church, Pinetops, NC

Traditional indoor venues for folk music have been cafes and similar small, intimate spaces. Although there are a few auditoria specifically intended for folk music, these are rare. Since folk music requires great intimacy, reverberation time in a venue designed for this music should be very low – perhaps 1.2 seconds. Intimacy, warmth, and clarity are important, as well as speech intelligibility and stage support. Acoustical strength is important as well, since folk music does not benefit from sounding "amplified"; preserving the natural sound of the instruments and voices is important.

The present increasing use of "contemporary" music in worship probably began in the "folk era" 1960's, as folk Masses appeared in some Catholic churches, and younger ministers of worship and music, and youth pastors, began to employ acoustic instruments other than piano and organ in their services. "Contemporary" worship music has tended to follow the stylistic trends of secular music into "pop" and rock styles, and today most contemporary worship music is of a style other than folk.

Rock

Figure 10-5: Sanctuary of Lake Forest Church, Huntersville, NC (Courtesy of Barry Parks, Thomas Hughes Architects, Winston-Salem, NC)

Rock music, as well as other styles of music depending heavily upon electronic amplification, is often fast-moving and highly percussive; therefore, it benefits from a very short RT60 – typically less than 1.2 seconds. This type of music is often played at very high sound levels, which command listener attention to the point that spaciousness and envelopment are of little importance. Since either in-ear or floor wedge monitors are usually provided for musicians (and sometimes for talkers as well), stage support is not very important. In fact, in order to avoid echoes from stage walls that can smear the sound of drums and cymbals, absorptive materials are often placed in the stage walls. Because of the rapidly-moving nature of the music, clarity and speech intelligibility are very important in rock venues. The high sound levels employed make HVAC noise less likely to be noticed than in other types of venues. The exception is any noise that has a steady-tone component, as these are particularly annoying and distracting even in the presence of loud music.

Rock style worship venues comprise those in which the style of worship music is rock, Nashville Gospel, and contemporary African-American Gospel. Rock-style worship spaces are distinguished from rock music venues by the fact that the programming includes speech and often drama as well. The requirements for good speech intelligibility are naturally met by the acoustical requirements of rock music. However, dramatic presentations, especially those involving amateur performers, benefit from good acoustical strength. Thus in cases in which performers are supposed to use microphones, but forget to do so or have equipment failures, or in cases in which the sound equipment is not sufficiently extensive for every performer to have a microphone (especially those in choruses), the natural acoustics of the room will assist in the program being heard and understood. Achieving good strength and low RT60 demands that acoustical absorption not be placed on the ceiling (a good rule, in general), and usually requires that pews or chairs have medium to heavy upholstery on both seats and backs. Usually, the wall at the back of the congregation will need to be quite absorptive to prevent echoes from the drums, and this will help to lower the RT60. If necessary, side walls near the back of the congregation may need to be absorptively treated as well. Any absorption used on stage walls to prevent percussion smear should be judiciously and sparingly placed.

A very common problem in rock venues, whether sacred or secular, is overpowering sound levels in the front most rows caused by the stage monitor speakers. The surest cure for this problem is the use of in-ear monitors, which are small receivers having ear buds, one of which is used by each performer. These allow the musicians and vocalists to have their monitors as loud as they want, without burying the audience in "monitor wash". In-ear monitors are also available that allow each user to have his/her own separate mix of instruments and/or voices in the monitor. Users of in-ear monitors should always be carefully educated in the dangers of overly loud sound, since these monitors are capable of causing permanent hearing damage.

In cases in which in-ear monitors are not feasible for budgetary or other reasons, often a physical rearrangement of musicians and vocalists on the stage can help reduce the needed loudness of monitor speakers. If separate amplifiers are used for electric instruments rather than all instruments being fed only through the sound system, the location of the amplifiers can be optimized to permit the use of lower monitor speaker levels.

Fig. 10-6: Large stage Monitor speaker to Provide Directivity (Courtesy Radian Audio, Inc.)

The third approach to reducing monitor wash is the use of directional monitor speakers. In this connection, remember that small speakers can control the directional radiation of sound only at high frequencies: a 6"-square horn speaker in a monitor can only control sound in the range above about 2 kHz (3 octaves above middle "C"). Thus to prevent excessive monitor wash, large monitor speakers with fairly large horns must be used: the Radian Audio Apex-1500 shown in Figure 10-6 is one example of an appropriate monitor speaker.

Speech

Venues intended for speech or non-musical dramatic presentations benefit from extremely short RT60. In addition, to enhance the naturalness of the speech, the reverberation should not vary much with frequency, meaning that acoustical absorbing materials must be carefully chosen with this goal in mind. These two requirements set a limit on the size of rooms that will provide a good acoustical environment for speech, both because larger rooms are more reverberant, and because in large rooms, the absorption of the air begins to be important due to the long path length for sound. Air absorption is effective mainly at high frequencies.

Meeting the two reverberation requirements also automatically provides excellent clarity, and excellent speech intelligibility as well, if the ambient noise is well-controlled. In speech rooms, whether lecture halls, council chambers, or "black-box" theaters, good acoustical strength is helpful, and is often best provided by reflective surfaces located close to the talkers. In some cases, low reverberation is provided by the use of a sound system incorporating flush-mounted ceiling speakers, along with an absorptive ceiling. In these cases, the sound system must compensate for the lack of acoustical strength.

Intimacy is another feature that is automatically provided by meeting the reverberation goals in a speech room. Envelopment and warmth are generally not important. Stage support is helpful in theaters and council chambers, though not important in lecture halls. As in any room requiring excellent speech intelligibility, speech rooms require very low ambient noise.

Multipurpose Venues

In fact, most auditoria and worship spaces must accommodate a variety of musical styles, along with a need for easily-understood speech. Therefore, it is common for multipurpose concert halls to either incorporate a form of variable acoustics (to be discussed in chapters 15-17) or to be designed for a compromise RT60 in the range between the Baroque optimum of 1.5 seconds and the Classical/Romantic optimum of 1.7-2 seconds. Often this compromise value is about 1.7-1.8 seconds for a concert hall, or about 1.5-1.6 seconds for worship spaces or multipurpose auditoria.

Worship spaces for "blended" worship often employ a range of musical styles ranging from folk music to Classical/Romantic. In some cases, rock music is used as well. Speech is always important in this style of worship. However, because of the wide range of variation from one such congregation to another, only very general statements can be made about the optimum acoustics; exact details must be worked out on an individual basis with a competent acoustician.

In general, the more heavily the musical emphasis relies upon electronic amplification and/or percussion instruments, the shorter the RT60 needs to be, down to about 1.2 seconds. The use of a choir and/or organ calls for a more reverberant space – up to about 1.7 seconds in a room for blended worship. In order to support amplified music or folk music well, clarity needs to be excellent. Good acoustical strength is always important in a worship space, because invariably there will be persons speaking or singing who do not use the sound system well, or

Figure 10-7: Christian Life Center of Elizabeth Baptist Church, Shelby, NC. (Photo by Steve Gardner, courtesy Elizabeth Baptist Church.)

at all. The more nearly the modern musical elements approach folk music, the greater the need for intimacy. Good stage support is also always important. Appropriate warmth is especially important when organ and choirs are used. Envelopment and spaciousness, while important for concert audiences, matter less in a worship space. And in any worship space, ambient noise must be minimized to improve speech intelligibility and to reduce distractions.

11. Preferred Acoustical Performance of Fellowship Halls, Gymtoria, and Life Centers

Many churches and conference centers have rooms variously designated as fellowship halls, multipurpose rooms, gymtoria (called by one consultant "gymna-café-churcha-toria", and by another, "multi-useless rooms"), parish halls, or life centers. These rooms always provide acoustical challenges for the designer and user, since they must by nature accommodate multiple uses. At one extreme, they are often used as dining halls, and as such have a need to permit privacy of conversation among table-mates and freedom from serving noises (dish clatter). This function indicates a low acoustical strength, and usually a low RT60. But they are also often used as lecture halls, which may need a higher acoustical strength, or as drama theaters, which definitely call for a higher acoustical strength. Still, dining, lecture, and drama all work well in a fairly dry (non-reverberant) environment. However, it is when such spaces are used for what we have called Classical musical performance or worship – with a corresponding need for a moderate RT60 – that the real design conflict becomes apparent. Table 11A illustrates some of the conflicts among the various normal uses of a multipurpose room.

84

Function	Noise	RT	Speech Intell.	Strength	Echoes
Meals	Low - moderate	Not critical, but enough absorption is needed to control sound level.	Low-Moderate	Low	None
Meetings	Low	Very short	Excellent	High	None
Worship	Low	Various – see Chapter 10	Excellent	Moderate-high*	None
Performances	Low	Short - moderate	Excellent	High	None
Play	Moderate	Moderate	N/A	**	None

Table 11A -- Acoustical Qualities Needed for Various Functions [Honeycutt]

*A room with high acoustical strength will encourage more enthusiastic hymn singing, but will also tend to emphasize foot-shuffling/coughing/sniffing noises.

**One factor that is not clear from this table relates to the type of play for which the room is used. If the room is to be used as a gym, most often in churches it will be for informal play, not for organized competition. However, in gyms used for organized competition, there is a widespread opinion that the sound of fans cheering boosts the athletes' performance; thus a room that reinforces the cheering (a "loud" room with high strength) would be preferred.

Figure 11-1 shows graphs of the RT *vs.* frequency for a room having dimensions of 120'X80'X36'. Bass frequencies are at the left of the graph, and high frequencies (consonants in speech, cymbals in music) are on the right. Graphs are shown for various conditions: unoccupied with gypsum ceiling, unoccupied with acoustical tile ceiling, and fully occupied as for a concert, with acoustical tile ceiling. Target RT values are also shown for various different room functions.

One way to approach the issue of acoustical strength is to designate a certain area of the room where the talkers or performers will be expected to be, and to provide surfaces to create strong early reflection in that area. This approach, while helpful,

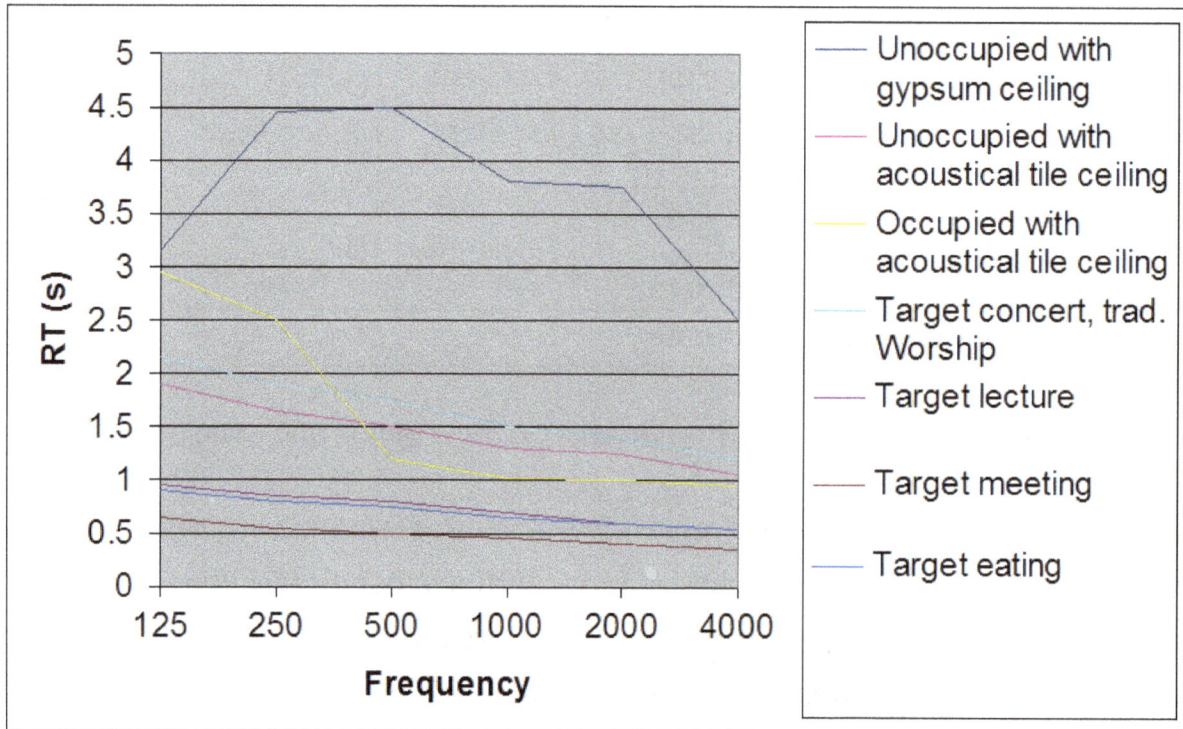

Figure 11-1: RT60 for Multipurpose Room [Honeycutt]

is usually not enough though, and reflective ceilings are desirable as well, except that these do not provide the conversational privacy desired by diners. A well-designed sound reinforcement system must usually come to the rescue.

The conflict in RT60 needs is more intractable, and can normally be solved only by the use of variable acoustics (see chapter 15) or the less-natural sounding expedient of using reverberation provided by a sound system.

Since multipurpose rooms are often adjacent to a kitchen, they are susceptible to a noise problem not found in other facilities; dish-handling and washing noises and food-preparation noises (e.g., blenders). The steel roll-up doors commonly used for serving in such facilities often do not provide a good noise seal, nor do the swinging entry doors, so there may be insufficient noise blockage from the kitchen to the listening space.

Some multipurpose rooms use a minimalist sort of design, with exposed HVAC equipment. Such an approach to HVAC design is susceptible to multiple mechanisms for producing excess noise, and should be carefully designed to prevent this problem.

12. Preferred Acoustical Performance of Viewing Rooms

Figure 12-1: The Extraordinaire Cinema at High Point University (Photo used by permission of High Point University).

Viewing rooms – once restricted to commercial motion-picture theaters – are appearing more and more frequently in educational institutions and worship centers as well. The audio portion of the movie experience changed from a monophonic track in the late 1920's to a stereo track, then a 3-channel track, and now to a multi-channel surround-sound track. For a mono sound track a typical speech room was quite adequate acoustically. For a stereo or 3-channel sound track, it became important for the viewer to be able to audibly determine the direction from which the sound came; otherwise, the stereo effect was lost. This required more careful design of the sound system and more careful acoustical design of the viewing room. Movies employing surround sound require extreme acoustical measures if the viewer's sense of aural involvement is to be convincing. In small rooms of about 1000 cubic feet, RT60 must be as short as 0.1 to 0.2 seconds, ranging up to no more than 1 to 2 seconds for very large rooms of 1 million cubic feet. Since the front right, left, and center speakers are pointed directly at the back wall, and they provide very high sound levels, freedom from echoes can require 4"-thick fiberglass treatment on the back wall. To keep the acoustical image from being modified by sidewall reflections, these are often treated with 1" or 2" fiberglass. Some acoustical warmth is permitted in viewing rooms, with the RT60 at 31.5 Hz allowed to be as much as twice that at 500 Hz.

Ambient noise levels in a well-designed viewing room should be as low as in a very good concert hall, requiring careful control of adjacent spaces and HVAC design. The sound of footsteps overhead or an air handler or elevator next-door will not enhance a viewing experience!

Since viewers will always compare their viewing experience to that in a good professionally designed theater, a private viewing room and its sound system should be designed with the aid of a consultant experienced in motion-picture theater design, and the consultant's involvement should commence during the architectural schematic design phase.

With these considerations in mind, it should be mentioned that not all viewing rooms must adhere to such strict standards. If the purpose of the room is informational or educational, the need for the extreme acoustic performance demanded by surround-sound theaters is not necessary, and a good speech room will work well. In such a case, most viewers are comfortable with a single speaker located above or below the center of the projection screen, and if the back-wall reflections are well-controlled and ambient noise is reasonably low, the room will perform satisfactorily. It is important, however, to remember that much of the information we receive from a video presentation comes to us aurally; thus, simply bringing a portable screen, projector and sound system into a gymnasium-like room is unlikely to lead to the desired results either in terms of entertainment or education. In worship facilities, a fellowship hall or multipurpose room may serve as an acceptable viewing room; a space designed for rock worship will be better.

13. Preferred Acoustical Performance of Rehearsal Rooms

Music rehearsal rooms, whether for vocal/choral music or for instrumental music, share several common needs that differentiate them from auditoria or worship spaces. The general concept of a music rehearsal room is probably that of a specialized classroom. However, it is more productive to think of in terms of a performance stage without the audience space attached. The first impression one may have is that a rehearsal room should sound as much like the performance venue as possible; however, this is a mistake. The RT60 of a performance space would not permit the clarity needed for the music director to hear individual performers in order to correct weaknesses. In fact, the much smaller enclosed volume of a rehearsal room compared to a performance space would make it impossible to achieve the same RT60 in both spaces, even if that were desirable.

A well-designed music rehearsal room should do seven things:
- Permit each performer to hear him/herself well.
- Permit each performer to hear the other performers well, to promote proper balance, intonation, and timing.

- Permit the director to hear each performer and the whole ensemble well.
- Keep the sound level comfortable even when the performance is fortissimo. (Low enough acoustical strength).
- Provide an RT60 that is comfortable for the performers, in terms of both length and frequency balance.
- Provide low enough ambient noise that the music is easy to hear.
- Provide enough sound isolation that rehearsals do not disturb occupants in other rooms in the building.

Achieving these objectives requires careful design that includes a knowledge of the expected size of the group to rehearse, as well as the flexibility to control room dimensions and finishes.

As in an auditorium, ceiling finish is critical in a rehearsal room. It must provide enough reflection of sound for proper "stage support", but enough absorption that the reflected sound is not too loud. It must also provide diffusion so that sound coming from various directions is reflected to each performer and to the director. Often these demands require that the ceiling be composed of a blend of reflecting, absorbing, and diffusing elements.

Walls are also an important source of significant reflections that must be controlled in order to provide optimum rehearsal conditions, and like ceilings, they also often require a combination of reflecting, absorbing, and diffusing elements. Unlike ceilings, walls provide an opportunity for the use of partially or fully filled bookcases or specific aesthetic elements to be used as diffusers.

Control of sound levels depends not only upon acoustical absorption, but also upon the distance from a performer to the nearest reflecting surface. Especially when risers are used, keeping performers far enough away from the ceiling requires a higher ceiling than the "standard" 8' to 9' height used in office spaces. When loudness control and the ability to hear "across the ensemble" (rather than only the adjacent performers) are factored in, the minimum ceiling height for an instrumental or choral rehearsal room is about 16'.

A value of RT60 that does not make performers feel as if they are performing on a large open field on a snowy day (very dead acoustically) is about 1 second for choral rooms, or 1.1-1.2 seconds for instrumental rooms. This value is difficult to achieve without proper ceiling height, especially in small rehearsal rooms, partially because RT60 depends directly upon the enclosed volume of the room, and partially because smaller rooms tend to be louder, requiring proportionally more acoustical absorption to provide loudness control. (Some people today are advocating very dead instrumental rooms in schools to control loudness and noting that the bands most commonly perform outdoors and the directors need to hear individuals who are learning. This plan would seem to apply mainly if the only instrumental group is a marching band, not a concert band, wind ensemble, or orchestra.)

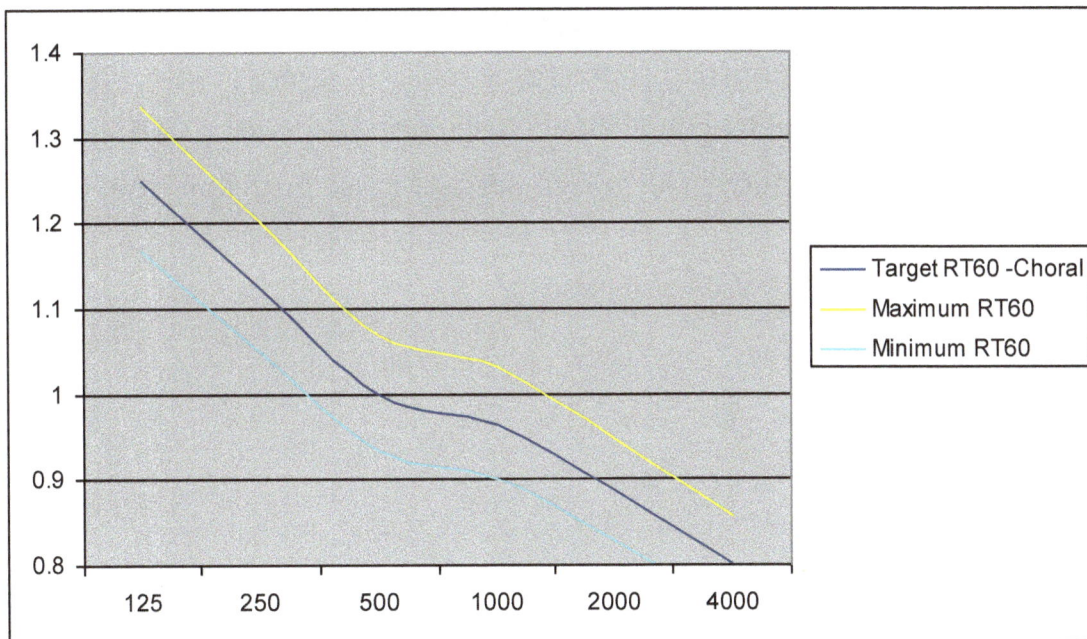

Figure 13-1: Target RT60 for Choral Rehearsal Rooms

Rehearsal rooms are more "music-friendly" if they have sufficient acoustical warmth. This points to an RT60 that is about 25% longer at 125 Hz (one octave below middle "C") than at 500 Hz (one octave above middle "C"), and about 20%

shorter at 4 kHz (4 octaves above middle "C"). Figure 13-1 illustrates target RT60 characteristics for typically-sized choral rehearsal rooms. For instrumental rehearsal rooms, the RT60 values would be increased by about 0.15 seconds. In general, carpeted floors provide too much high-frequency absorption, making the room sound dull, and preventing proper diffusion of sound, so hard floors are preferred for rehearsal rooms. The possible exception is that risers can be lightly carpeted, since they will be occupied by absorptive people, blocking most sound from hitting the carpet.

In addition to melody, harmony, expression, and balance, a music director must also pay close attention to articulation: the beginning and ending of notes, and the diction. The ability to do this well requires freedom from acoustical distractions in the form of HVAC noise or noise entering the room from adjacent spaces. As mentioned earlier in this book, control of HVAC noise is a matter of system design, while isolation of external noise (and prevention of rehearsal sounds from disturbing occupants of adjacent rooms) depends upon the construction materials used. Massive materials such as solid concrete, or multiple layers of materials such as gypsum board, can provide good sound isolation. Suspended ceilings may require a barrier of gypsum above the grid to prevent noise intrusion from rooms above.

V. Controlled Acoustics

14.   Sound Systems

Intro to Sound Systems

Ever since the ancient Greeks introduced resonators to improve their theaters, efforts have been made to control the acoustics of spaces for performance, worship, and speaking/hearing. Undoubtedly the most common method in use today is sound reinforcement systems. Just as there are many different purposes for listening spaces, so there are many different applications for sound reinforcement systems. This chapter will address some of the design targets and parameters common to sound reinforcement systems.

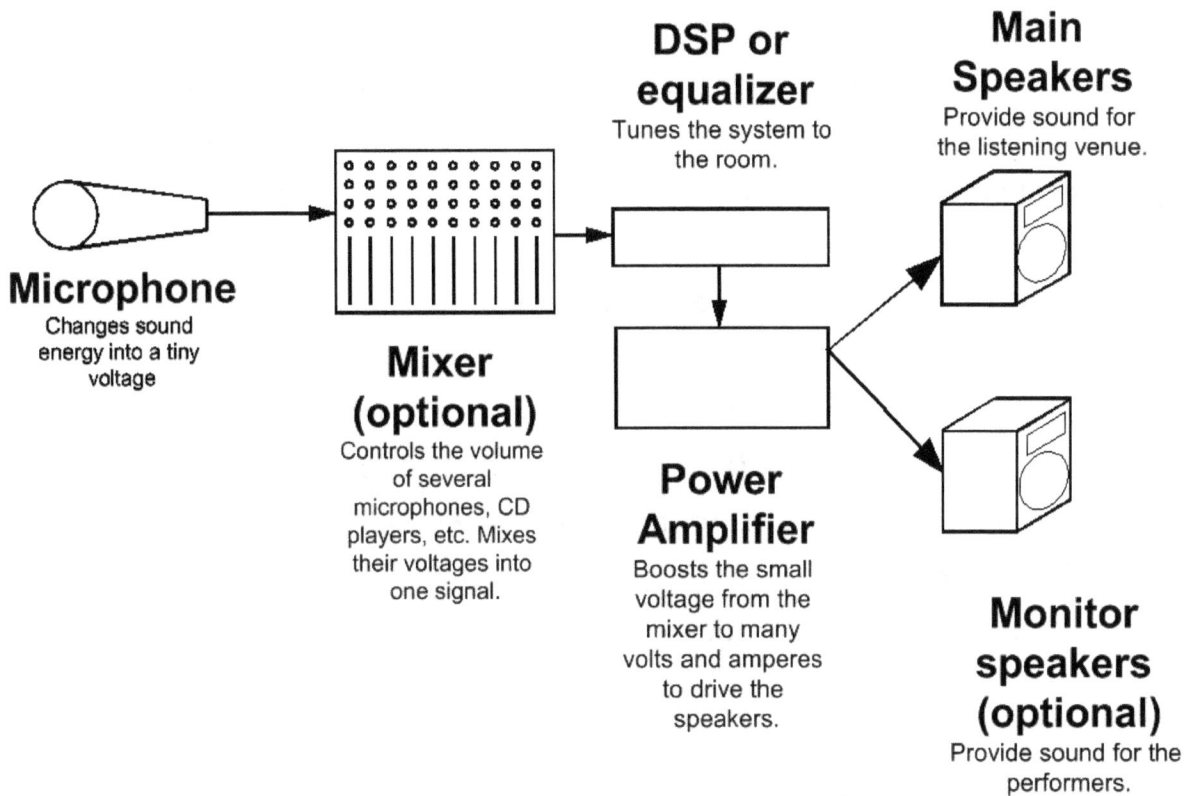

Figure 14-1: Block Diagram of a Basic Sound System

As illustrated in Figure 14-1, all sound reinforcement systems include at least three elements: a source, one or more amplifiers, and speakers. The source may be a microphone, which changes sound into electrical energy; it may be a CD, DVD, minidisk player, computer, MP3 player, or tape player. An amplifier increases the power of the electrical signal from the microphone or other source from microwatts to tens, hundreds, or thousands of watts of electrical power. The speaker(s) change the amplified electrical power back into sound. It is important not to confuse microphones with loudspeakers, if clear discussions about sound systems are desired. Remember: microphones pick up sound; speakers make sound.

Most sound systems also include a mixer to allow selection and control of several sound sources. Many also include a digital signal processor (DSP) or an equalizer to permit fine control of sound quality or timbre. The equalizer is often a part of a digital signal processor (DSP) that may also include other functions.

Design Targets -- Speakers and Amplifiers

The first target for any sound reinforcement system in a performance or worship space is that the system provide the right sound level to listeners. This can vary from about 50 dBSPL for background music to well above 100 dBSPL for rock venues. The target sound level affects many aspects of system design, and it is therefore vital for the system designer to know accurately what functions are expected of the system before work commences.

A parameter related to the sound level for any sound system using microphones (rather than only prerecorded sound sources) is the **needed acoustic gain (NAG)**. This is a measure of how much the sound from the microphone has to be boosted in order to make the sound from the sound system at a listener's ear as loud as it would be at a specified distance from the talker without the system. Thus, NAG depends upon the distances from the talker to the microphone, and from the talker to the listener, as well as the chosen reference distance. For example, if a person talking in a loud voice produces 70 dB at a microphone 2' away, and someone else listens from 8' away, while a third person listens with the help of a sound system from 128' away, then in order for the sound levels to be equal for the two listeners, the NAG is 30 dB, which allows a 6 dB margin for stability against feedback squeal.

(Feedback squeal is caused by sound from the speaker getting back into the microphone and being repeatedly re-amplified.)

Of course, in the real world, there are limitations on what a sound system can do. Thus there is also a parameter called the **potential acoustic gain** (**PAG**). This is the maximum gain that a system can provide in a given space, based upon the parameters of the space and the system. PAG is ultimately limited by the need for stability against feedback squeal. Given the situation just described, if the speaker is 45' above the microphone and the listener is 90' from the talker, the PAG is 30 dB. If this system had to be turned up to provide more gain to accommodate a weak talker, it would likely squeal. (These calculations are simplified by assuming outdoor conditions (no reverberation); a system designer would use a more accurate approach that takes into account the acoustics of the room.)

The above calculation for PAG assumes that only one microphone is used at a time, and that the microphone and speaker are not directional. If more than one mic is used at a time, each doubling of the number of open microphones reduces PAG by 3 dB which means in a real situation that it increases the likelihood of feedback squeal. Using directional microphones and speakers increases PAG. If the talker is closer to the microphone, as with a properly worn lavaliere mic – or better yet, an earset mic – the PAG is increased.

The second target is that all listening areas receive as nearly as possible equal support from the system. Those seated close to the front should not be penalized by uncomfortably loud sound, nor should those in the rear have difficulty hearing. Two ways of attaining even sound distribution are via **distributed** speaker systems and by **central array** systems, although there is a hybrid called the **exploded** or **distributed array** that is sometimes used as well.

A distributed speaker system consists of many separate speakers, each of which covers only a small portion of the listening area. Typical of distributed systems is an installation having flush-mounted ceiling speakers. These systems are generally inexpensive and can work well for low to moderate sound levels in spaces having ceilings 8' to 12' high. In order to adequately control the sound radiation, higher ceilings will call for larger speakers. Ceiling-mounted speakers are not suitable when sound levels above about 90 dB are needed, or where extended bass response is desired. They are inherently limited by the fact that if the budget for speakers is split among dozens of units rather than just a few, as in array designs, each speaker must necessarily cost less and therefore will not be of as high quality. Also, any attempt to achieve high sound levels, especially at bass frequencies, may excite rattles and buzzes in the ceiling system. Another inherent limitation of pure distributed speaker systems is that the "acoustical image", or place from which the sound seems to originate, is not the same location as that of the talker or singer. This confusion of acoustical imaging can be distracting for listeners. (It is possible to create some acoustical imaging in distributed systems, using a combination of delay and level control of groups of speakers, but this involves a complex design and set-up process and is not often done.) With these caveats, it is fair to say that distributed ceiling speakers are a cost-effective solution for speech and background music if their limitations are kept in mind.

Large listening areas such as sports venues sometimes utilize distributed speaker systems, although in this case, the individual speakers are usually higher-quality enclosed systems that can provide a wider frequency range and higher levels than can ceiling speakers.

Where distributed speaker systems are not appropriate, the best solution is usually a central array. This may consist of only one speaker (in which case it is "central" but not an "array"). Sometimes a building owner will prefer the use of a speaker on each side of the room, which can work reasonably well, except for one problem: acoustical imaging. For listeners on either side of the room and located close to a speaker, the acoustical image will be at the nearest speaker. A central array avoids this problem for the most part, since people are much more sensitive to the horizontal position of a sound source than to the vertical position. Thus a centrally-mounted speaker will produce an acoustical image much closer to the talker/singer location than will side-mounted speakers. It is possible, however, for a central speaker to be mounted so high that a "voice of God" effect is produced because the listeners localize the sound well above the talker's position. The use of small speakers mounted under the stage lip, or in platform stair step risers in a worship space, can bring the acoustical image down and provide a more natural listening experience.

Another advantage of a central array is that it naturally provides more even sound coverage from front to back than side-mounted speakers. This is true because sound level decreases as a function of distance from speaker to listener. If a speaker is mounted at a height of 30' in a room 90' from front to back, and the listener area begins directly under the speaker, the distance from speaker to farthest listener is only 3.6 times the distance from the nearest listener; whereas if the speaker is mounted 8' high on a wall 4' from the front row of listeners, the furthest listener is 22.5 times as far away as the nearest. This translates to a front-to back variation in sound level of 13.5 dB for the latter case as compared to 5.5 dB for the former.

Figure 14-2: Renkus-Heinz Model STX2 Speaker

In rooms having fairly low ceilings, but using music that has a level above 90 dB, and/or is rich in bass content, neither a pure distributed system nor a single central system is suitable. In this case, an exploded array (also called a distributed array or virtual array) can be used. This type of system typically uses one or more main speakers located close to the stage, with "fill speakers" located farther back. The location and directivity of the fill speakers is selected to provide even sound coverage at all frequencies, and the signals to the fill speakers are electronically delayed so that the sound from those speakers will be synchronized with the sound from the front speakers, which is acoustically delayed by about 0.88 milliseconds per foot. (In an actual installation, the fill speakers' signals may be delayed by a few milliseconds more than the acoustical delay, to permit the precedence effect to fool listeners into perceiving the sound as coming from the front of the room, rather than from the fill speakers.)

The types of speakers appropriate for sound reinforcement arrays include directional multi-way systems that normally use cone-type low-frequency speakers and high-frequency horns (see Figure 14-2), coaxial systems in which the cone of the low-frequency speaker acts as the walls of the high-frequency horn, and line arrays consisting of many individual single or multi-way speaker units arranged in a vertical line. The "sound columns" of the mid-twentieth century were primitive line arrays. The first two types of speaker can be called "single-point systems", since the sound emerges from a small region in space (not truly a point) and spreads as a section of a sphere. This type of speaker transmits sound equally in all directions at low frequencies, up to about 500 Hz (an octave above middle "C"). At higher frequencies, it becomes more directional, and at the frequency at which the horn takes over (varies among different models), the directivity becomes more-or-less constant, and is specified in degrees of horizontal and vertical coverage. If the horn is circular, the horizontal and vertical coverage angles are the same. Rectangular horns can have different horizontal and vertical coverage angles. In all actual horns, the coverage angles vary somewhat with frequency, being much larger at low frequencies and narrowing to the specified angles (perhaps 90° horizontal by 60° vertical) at higher frequencies. In some cases, the coverage angle actually drops below the specified value at frequencies of a few kHz. For all these reasons, sound

Figure 14-3 SLS Model LS8695v2
(Photo courtesy of Renkus-Heinz)

system designers have to use computer modeling software that calculates coverage based upon the measured sound radiation of the speakers, rather than depending upon specified coverage angles.

Modern line arrays come in two flavors: non-steerable (Fig. 14-3) and electronically steerable (Fig. 14-4 and 14-5). A non-steerable line array radiates sound as a section of a "flying saucer" whose height equals that of the array, for locations near the speaker, but tapers to a narrow beam farther away. This requires the array to be mounted at the listeners' ear height, or to be tilted forward if mounted higher. The horizontal coverage angle of most line arrays is about 120° to 150°, although one company produces a "touring" line array (vertical row of large speaker boxes) that has selectable horizontal coverage. Line arrays have advantages and disadvantages when compared to point-source speakers:

• The horizontal sound coverage of a line array is very uniform and wide; whereas point-source speakers seldom can boast 120° of horizontal coverage. However, this can be a disadvantage in reverberant rooms where it is desirable to keep as much sound as possible off the walls to improve intelligibility and clarity.

• The frequency response (sound output variation with frequency) of a line array speaker varies with distance, so designing systems using them optimally is not a trivial exercise. A perfect speaker's output would not vary with

Figure 14-4: Renkus-Heinz Model IC16R
(Photo courtesy of Renkus-Heinz).

frequency; such a speaker would be said to have "flat" frequency response.

- Because of their usual placement close to the listening plane, line arrays are more likely to be near listeners than are typically higher-mounted point-source speakers, resulting in possibly uncomfortably loud sound for those nearby listeners.

- The ear-level mounting can create vulnerability to echoes from the rear wall.

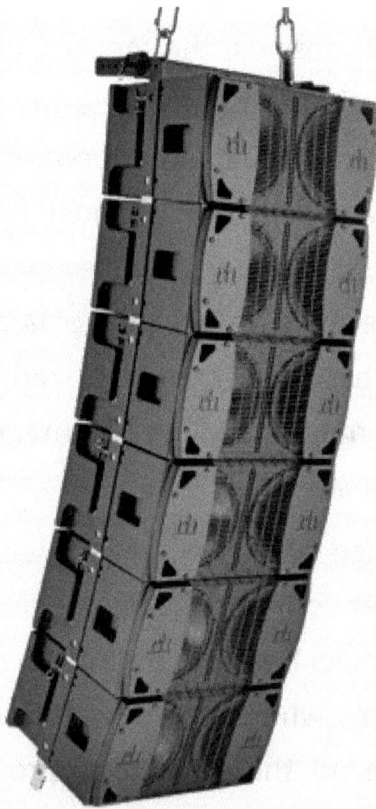

Figure 14-5: Renkus-Heinz PNX Line Array Speaker (Photo courtesy of Renkus-Heinz).

Steerable line arrays have a separate amplifier and digital signal processor for each element (speaker or horn/speaker pair). They can point the beam at different angles vertically, rather than only straight out, as non-steerable arrays do. Some models can even provide multiple beams so that, for example, one or two beams could be pointed toward the rear seats and operated at a higher sound level in order to overcome the decrease in level with increasing speaker-to-listener distance. Even with these performance advantages, the comparison of line array versus point-source speakers given above still applies. A competent, experienced sound system designer will be able to select the correct type of speaker system for a given venue.

Steerable line arrays are available in two varieties: single-box systems as shown in Figure 14-4, and concert line arrays as shown in Figure 14-5. The concert arrays can provide excellent control of sound radiation at greater distances than can the single-box systems, and can also provide more acoustic output (greater sound levels). In order to provide these performance advantages, concert line arrays are much larger than are single-box systems.

Another sound system design target is related to evenness of coverage. This is the need to avoid excessively exciting the reverberant field in the room. Basically, this means "put the sound where the people are". Sound that strikes the listener areas is useful, and also gets absorbed by the people. Sound that misses the listeners and hits the walls or ceiling bounces around the room, adding to reverberation, decreasing clarity and speech intelligibility. Thus most venues need speakers having very specific horizontal and vertical coverage angles. This requirement is not met by most portable "speaker on a stick" systems. Much advertising has been done promoting the use of steerable line arrays that can offer beams as narrow as 5° in vertical angle. This will indeed help keep excess sound off the ceilings. However, with a 120° to 150° horizontal coverage, a lot of sound will hit the walls. Thus the choice of which type of speaker will perform more suitably in a given venue depends upon the shape and composition of the walls and ceiling, as well as the shape and orientation of the listener areas.

Based upon the PAG and the desired sound level of any amplified music, the sound system designer can choose which particular model of speaker will provide the needed sound level and will evenly cover the listener area. After selecting speaker model and location, the remaining decision to be made regarding providing sufficient sound level involves choice of amplifier power. Knowing the specifications for the chosen speaker(s), the designer can determine how much power is required to drive the speaker(s) in order to produce the desired level, including a 10-to-20-dB factor (multiplier of 10 to 100) for sound peaks. While the speech-and-vocal-solo sound systems of the mid-20th century often used amplifiers of 100 watts or less, meeting the demands of today's music especially in large venues often requires amplifiers having output powers in the thousands of watts. It is not unusual for a sound technician to receive a complaint of a failed speaker, only to discover that the amplifiers were being overdriven in an attempt to get enough sound level from an underpowered system. Overdriven amplifiers sound remarkably like damaged speakers, especially to the untrained ear.

Design Targets – Electronics

Now we finally get to the design targets that most users think of first: usability considerations. These include the following:

- How many input sources are needed? These include wired microphones, wireless microphones, tape, CD, and DVD players, audio from computers, and any other source of audio program material.

- Will a person be assigned the task of operating the system for every event?

- Will the system use more than one type of speaker, or speakers in more than one location?

- Will stage monitor speakers be needed?

- How many, if any, wireless microphones are needed?

- Does the system need to provide audio and/or video recording capabilities?

- Would it be advantageous for the system to have programmable, memorizable control settings?

- Will there be any instances in which the system will be called upon to serve as a portable system, or will it be permanently installed?

Mixers – Analog

As shown in Fig. 14-1, the device used to control the levels and timbre of individual sources is called a **mixer**. These are available in a bewildering variety of models. Simple systems having eight or fewer sources can sometimes use a powered mixer or mixer-amplifier such as the one shown in Fig. 14-6. This particular unit includes 12 sets of controls for sources, two of which accommodate stereo sources; a graphic equalizer for controlling the timbre of the whole system; a compressor (keeps loud signals from getting too loud); and a dual 500-watt power amplifier.

Figure14-6: Yamaha EM-5016CF Powered Mixer (Photo coutesy of Yamaha).

Lower-budget installations that do not require as much control can often use a simpler mixer-amplifier such as the one of those shown in Fig. 14-7. These provide much less control than the unit shown in Fig. 14-6, and are available with a single power amplifier of up to 160 watts.

Figure 14-7 Crown MA Mixer-Amplifiers (Courtesy of Crown International.)

Many sound systems require more input channels than are available in a powered mixer, or require more flexibility than a single-unit system can provide. (A **channel** is a set of controls and associated electronics that affect a single input source.) For these systems, the mixer, equalizer, compressors (if any), and other signal processing devices are separate devices. It is not unusual for a well-equipped drama theater or house of worship to need a mixer having 48 or more channels; indeed, mixers are available having well over 100 channels. Figure 14-8 shows a 48-channel analog audio mixer.

Which particular mixer a sound system designer chooses involves more than the number of input channels. If stage monitors are to be used, separate "auxiliary" mixes should be provided for these. In many applications, it is desirable to be able to group channels together so their levels can all be controlled at once; for example, if a choir is amplified using multiple microphones. In this case, a mixer having enough subgroups is helpful. Since pure stereo sound reinforcement systems are not popular for live sound reinforcement (people seated on house-right tend to miss what's happening in the left channel, and *vice-versa*), a mono output may be helpful. On the other hand, where a stereo or a left-center-right system is used, the mixer needs to have the appropriate outputs and controls to accommodate this arrangement. If programs are to be recorded directly onto a computer, a mixer with a USB (universal serial bus) output is a good choice. Also, the controls available on each channel vary from model to model. These include equalization, which may range from simple 2-band equalizers to more extensive 4- or more-band "sweepable" equalizers, and compression. Many of the other fine points of mixer selection require experience with sound-system operation to even be understood, so we will move on to other equipment.

Figure 14-8: Allen & Heath Model GL2800-48 Mixer
"Photo courtesy of Allen & Heath"

Regardless of the mixer chosen, the physical location of the mixer in the room is of critical importance. In order to accurately control the sound in a room the system

Figure 14-9: Yamaha Model 2031B Graphic Equalizer (Photo courtesy of Yamaha Corporation).

operator needs to hear sound that represents what the audience or congregation is hearing. The sound quality is different at different places in the room: in a balcony, the reverberation may be less pronounced; whereas under a balcony, both direct and reverberant sound are usually not as loud. Some venues locate their sound controls in a walled-in room or booth, perhaps with a small window opening into the main room. This is like asking a person to drive a car from the back seat. If a calibrated monitor speaker system is installed in such a room, and if people can be restrained from fiddling with the controls, at best barely tolerable results can be obtained from such an arrangement. However, really good sound demands that the operator be located where (s)he can hear sound typical of what the audience/congregation hears. Where visual aesthetics triumph over audio function in the chosen control location, less-than-optimum results must be expected.

Equalizers and DSP's

Since no speaker is perfectly accurate, most sound systems include some means of leveling out the frequency response of the whole system. If more than one type of speaker is used, this becomes a necessity. The time-honored piece of equipment for this task is the graphic equalizer, such as the one shown in Fig. 14-9. Graphic equalizers break the signals into different frequency bands and provide control of

Figure 14-9: Yamaha Model 2031B Graphic Equalizer (Photo Courtesy of Yamaha Corporation)

each band. Although octave-band (also called 8-band or 9-band) and 2/3-octave-band (also called 15-band) equalizers are available, for serious control of frequency response irregularities, a 1/3-octave-band (30 to 33-band) equalizer is preferred.

A frequent problem with graphic equalizers is tampering by people who have neither the knowledge nor the measuring equipment to adjust them properly. Digital signal processors (DSP's) provide an alternative method of equalizing sound systems, while providing additional features as well. As an added bonus, these are much less subject to tampering, since their controls must be accessed through computer interfaces requiring specialized software. Not only can a DSP provide the functionality of a graphic equalizer, but many models can provide a more precise equalization function called a **parametric equalizer**. They can precisely delay the signals for rear speakers so that the sound from them is synchronized with that from the main speaker. Many DSP's have multiple inputs and outputs, with equalization, delay, and compression available separately on each input and each output. Also commonly included is an **auto-mixing** function that turns microphone channels on and off according to the sound level sensed at the microphone or other source. For systems not having an operator, this function allows the use of multiple microphones without engendering feedback squeal caused by having all the mics turned on at the same time. Physically, a DSP is a rather plain little box – not much to look at. All the exciting stuff is on the inside.

Digital Mixers

For venues that need programmable mixer setups, a digital mixer may be a viable alternative. Although much more expensive than an analog mixer, a digital mixer includes many of the functions of a DSP plus an outboard effects processor for special reverb or echo effects. Using a digital mixer, the sound technician for a drama theater in which a certain play is rehearsed one night a week for a month before the performance could set up the mixer for each scene in the play, then store the settings in the mixer's digital memory. Then after the first rehearsal, the settings for each scene could be recalled in order with the push of a button. While similar functionality is available in analog mixers having motorized faders (controls), the automated control functions of a digital mixer are much more extensive. In addition, digital mixers provide much greater control over each input channel and each output as well. Fig. 14-10 shows a digital mixer.

Figure 14-10: Yamaha M7CL48ES Digital Mixer (Photo Courtesy of Yamaha Corporation)

This particular mixer also has a stage box for microphone connections in which the mic signals are converted to digital format and then sent to the mixer itself through computer networking cable, making long runs of 96-conductor shielded copper cable unnecessary. Notice the upward-sloped section to the right of the mixer's center. This is called the **control surface**, and contains the LCD display and knobs through which most control is performed. Many digital mixers have more than one **layer** of controls; for example, a 48-channel mixer like the one shown might show 16 channels of controls on layer 1, another 16 on layer 2, and the rest on layer 3. This allows sufficient space for viewing and operating controls for 48 channels while keeping the size of the LCD panel (as well as the mixer itself) reasonable. The main disadvantage of a digital mixer other than the price is its complexity of operation. If the venue employs a full-time audio technician, this complexity should not present a problem, but if reliance is on volunteer technicians, the steep learning curve can be problematic.

Some newer digital mixers offer a feature that simplifies the problem of control location: they can be controlled by a portable computer running specialized software and connected through a wireless network to the mixer. This feature allows the actual mixer to be completely out of sight, while the operator can be anywhere in the venue.

Microphones – Wired and Wireless

Many types of microphones have been invented since Alexander Graham Bell developed the first one in 1875. Those in common use in sound systems have either an electromagnetic (or **dynamic**) principle of operation, or an electrostatic (or **capacitor**) principle. Without getting into the fine points of microphone choices and techniques, on which many books have been written, suffice it to say that capacitor microphones require a DC voltage to be supplied to the mic, while dynamic mics do not. In most portable devices, the voltage is supplied by internal batteries; whereas, in sound systems, the voltage is supplied by the mixer through a system that uses

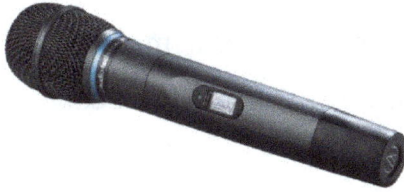

Figure 14-11: Audio-Technica Handheld Transmitter (Photo Courtesy of Audio-Techinca.)

Figure 14-12: Lectrosonics Beltpack Transmitter (Photo Courtesy of Lectrosonics, Inc.)

the mic's signal cable to supply the voltage. This method of supplying voltages is called **phantom power**. Not all mixers supply phantom power, and those that do not will not work with capacitor mics unless an external phantom power adapter is used. Some mixer that supply phantom power have a single switch for turning the phantom power on or off, and others have separate switches for each channel or group of channels.

Many, if not most, sound systems installed since the 1980's include wireless microphones. A wireless microphone is a system involving a small portable radio transmitter and a separate small radio receiver. The transmitter comes in three forms, a belt pack unit into which a small lavaliere (clip-on) or earset (also called "earworn") mic's cable can be plugged, a unified handheld mic/transmitter, and a plug-on transmitter designed to be used with a separate non-wireless microphone. The receiver is usually located near the mixer unless the mixer is

hundreds of feet or more from the stage. Wireless mic receivers come in a variety of designs and price ranges. Most good wireless mics use the **diversity principle**, in which the receiver is actually two receivers, each with its own amplifier. Special circuitry in the receiver selects the one providing the best signal, thus reducing interference

Figure 14-13: Lectrosonics Plug-On Transmitter (Photo Courtesy of Lectrosonics, Inc.)

and "dropouts" (moments of signal loss, usually accompanied by a loud hiss). The better units also use either **tone squelch** or **digital code squelch** to mute the receiver when the transmitter is turned off or out of range, thus preventing noise bursts in the sound system.

Most manufacturers of wireless microphones can provide various types of antennas, from the small stick-like "whip antenna" through several models of directional antennas that can extend the range (maximum transmitter-to-receiver distance) of the wireless system.

Fig. 14-14: Lavalier (left) and earset (right) microphones (Photos Courtesy of Audio-

There are basically three microphone types that can be used with a belt pack transmitter: a non-directional lavaliere (clip-on), a directional lavaliere, or an earset mic. Non-directional lavalieres are usually best for non-professional users, as directional ones require careful placement and orientation in order to work well. Earset microphones place the mic closest to the talker's mouth, providing the best immunity from feedback squeal. Since the mic is beside the mouth, there is no problem with "p-popping". Figure 14-14 shows a lavaliere and an earset microphone.

A wireless mic receiver can only be used with one transmitter at a time. Each transmitter and receiver operates at a specific frequency, and if two transmitters are operating at the same frequency, the receiver will only pick up the one whose signal is stronger. This can result in very unpleasant interference if the transmitters are about the same distance from the receiver. Some wireless systems are sold as a 2-transmitter (handheld, belt pack) – one receiver system, but these are intended for use with only one transmitter active at a time.

One of the challenges of successful wireless microphone use is frequency coordination: making sure the systems do not interfere with each other or with other radio services. The Federal Communications Commission (FCC) is responsible for deciding what type of radio, TV, or other wireless service can use which frequencies, and the rules change from time to time. Wireless mics – except those used by radio and TV broadcasters – do not require a license, and unlicensed services are required by the FCC to shut down if they interfere with licensed services, but the reverse is not true. Therefore, if a wireless system in a performance or worship venue is subject to interference, the interfering service usually cannot be required to cease operation. The exception is one group of wireless devices for home use, which incorporate circuitry to force them to change operating frequency to avoid interfering with any other service. As to wireless mic systems interfering with each other, this can be a problem within a single facility if the various systems' frequencies are not properly coordinated. Also, it can be a problem when adjacent facilities both use wireless systems. The higher models of most wireless mic manufacturers' lines include scanning devices to locate frequencies clear of possible interference. Especially in large cities, the extra money for such a system is well-spent.

Hearing Assistance Systems

Performance and worship venues are not places in which it is easy for hearing-challenged people to hear and understand programs. Even with low ambient noise and the best possible clarity and speech intelligibility, people with significant hearing disabilities will have difficulties. A way to help these people involves an electronic hearing assistance system. Any facility serving hearing-challenged people should use such a system, and some facilities are required by law (the Americans with Disabilities Act of 1990 and subsequent amendments) to provide hearing assistance devices to any patrons who request them.

These devices come in three varieties. The more common uses a small FM radio transmitter fed from the sound system, and a small FM radio receiver for each listener. In some venues, it has been found that pirate recordings of concerts and other programs were being made by persons located near the venue, by receiving and recording the signals from these FM hearing assistance transmitters. Another type of system is therefore favored by such venues: one using infrared (IR) light rather than radio. With careful placement of the IR transmitters, listeners find these to perform as well as the FM systems, but since the IR light is confined by the room, the signals cannot be picked up from outside for surreptitious recording. An added advantage is that IR systems do not require frequency coordination as do FM systems. Although there are a number of different frequencies on which the FM hearing assistance systems can operate, finding an available frequency in a congested theater district of a large city can be difficult.

The third type of system, called an **inductive loop** system, employs a large wire loop encircling the audience area. If listeners within the loop have hearing aids with "telecoil pickups" (special means used to help hearing-impaired people hear telephone conversations, available from hearing-aid suppliers), they will be able to hear the signal from the sound system. Telecoil receivers are available for listeners not having telecoil-equipped hearing aids. Loop systems are also immune to concerns about piracy or interference with other venues or communication services. These systems require careful engineering to avoid hums and buzzes from other electrical equipment in the building.

The user of a hearing assistance system can hear the sound through either of three mechanisms, as shown in Figure 14-15. The most common is the ear bud, which is placed in the user's ear canal and its sound is therefore nearly inaudible to those seated nearby. There are sanitation issues associated with ear buds that are used by multiple listeners, however, and even though they are covered with disposable foam sheaths, replacing the sheaths becomes a maintenance issue that all too often is not remembered. The second option is the ear speaker, a small device that clips around the user's outer ear and places the speaker close to the entrance of the ear canal. Since no part of the ear speaker actually enters the ear canal, sanitation issues are much reduced, and some people find ear speakers more comfortable than ear buds. Users having hearing aids equipped with induction coils to better pick up sounds from telephones have a third option: a neck loop. This is a flexible wire "necklace" that can be plugged into the hearing assistance receiver instead of an ear bud or ear speaker, and that will be picked up by the hearing aid. This option takes advantage of the personalized settings available on the best modern hearing aids, and thus are best for those having severe hearing impairment.

Portable Sound Systems

A full discussion of the many choices in portable sound systems is beyond the scope of this book, but a few comments are in order concerning their use. First, a portable sound system is subject to a number of compromises: the speakers must be small enough to be easily carried; the speaker locations are restricted to fairly small heights; and the electronics must be chosen with a view toward portability. The exceptions to these generalizations are the touring sound systems that are used for large concerts and travel in semi-trailers.

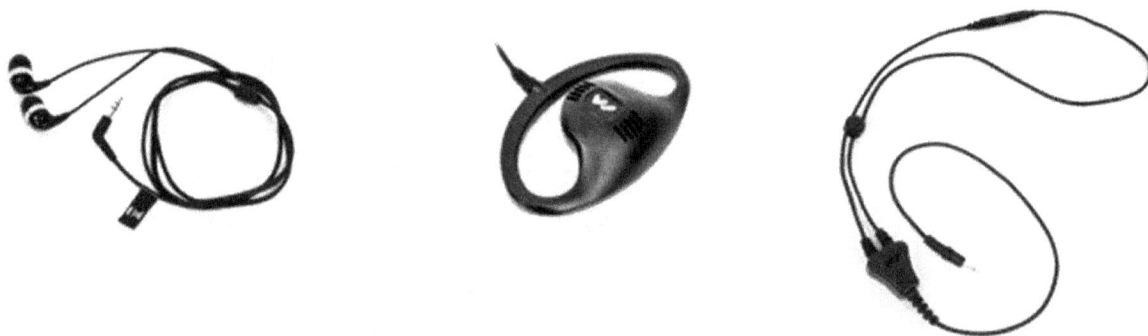

Figure 14-15: Earbud, Ear Speaker, and Neckloop (left to right) for Hearing Assistance (Photo courtesy of Williams Sound)

Typical portable sound systems are properly employed to cover a fairly small group of people out-of-doors or in a room having no permanent system. When used as a long-term substitute for an installed system, the shortcomings mentioned above will limit the performance of the system. Also, the careful speaker placement required to provide even coverage and maximum PAG is generally not possible with typical portable systems. Indeed, the directional performance of most small portable speakers varies so much with frequency as to make *any* placement a compromise.

Stage Monitors

While performers of unamplified music must rely on the stage support of the venue in order to be able to hear themselves and each other, performers using amplified instruments or voices commonly use stage monitor systems. Indeed, even some pastors prefer to hear themselves on a stage monitor while they preach. Simple stage monitor systems feed the same signal (or "mix) to the monitors that is fed to the main speakers. More sophisticated monitor systems allow the monitor mix to include only certain sources, with the levels of those sources controlled independently. For example, this allows the monitor system for a choir to exclude suspended microphones that are used to pick up the choir, so as to reduce problems with feedback squeal. The monitors in this case are used to allow the choir to hear the instrumental accompaniment as well as other parts of the program (sermons, announcements, etc.).

Stage monitor speakers may be roughly classed in one of four varieties. Floor monitor speakers (often wedge-shaped; hence, the moniker "wedge monitors") serve a small group of performers. "Side-fill" monitors can serve all the performers on a stage, if high sound levels are not required from the monitors. Overhead monitors attempt to serve the same purpose as side-fills. Personal monitors, often mounted on mic stands, serve a single person. In-ear monitors also serve only one person. It is important at this point to distinguish between a stage monitor and a studio monitor. The latter is designed to provide an accurate means of hearing what is being broadcast by a radio or TV studio, or what has been recorded by a recording studio. These units are not designed to be rugged enough to survive long as stage monitors, nor do they typically have stable enough directional characteristics to minimize feedback squeal. In many cases, they are not capable of sufficient output power to be heard well in stage.

As mentioned in chapter 10, in order for floor monitors to avoid producing objectionable levels of leakage (monitor wash) into the audience area, they must be of substantial size, usually employing at least a 12" cone speaker and a sizeable high-frequency horn. Where stage monitors are used, all microphones (with the possible exception of lavaliere and earset mics) should be directional, and the microphones should never be pointed into the monitor speakers. All of this means that the microphone should be in front of a vocalist, pointed at the vocalist's mouth, and the floor monitor should be on the floor farther in front of the vocalist, and pointed back at the vocalist. It is good to remember that a floor monitor can cause feedback squeal because of sound bouncing off a reflective wall behind the performers.

Side-fill monitors are often inset into walls beside the stage or platform area, and their directionality is not as much of an issue as is that of floor monitor speakers – the reason being simply that side-fill monitors are generally used at lower levels, or in some cases, only to enable singers to hear their accompaniment. Thus monitor wash from side-fill monitors is less a problem than with floor monitors. In many cases, a small non-steerable line array works well as a side-fill monitor.

Overhead monitors are used when there is no other choice. As a group, they create even greater problems with monitor wash than do floor monitors, because, being located far from the performers, they must produce high sound levels to be heard well. The conspicuousness of ceiling location often prohibits the use of large enough speakers for good directivity, and bounce off the wall behind the performers is also a frequent problem.

The small size of personal monitors prohibits significant directionality except at rather high frequencies, and also generally imposes limitations on smoothness of frequency response. They are useful, for example, in cases when an organist and a pianist play together but are physically separated by a large distance, yet need to be able to hear each other without significant delay. (The delay encountered by a sound wave traveling 71' corresponds to the length of a 32nd note at 120 beats per minute.) In this case, the monitor speaker is nowhere near the mic that provides its signal, yet it is near the performer who needs to hear it, so its level need not be very high.

In-ear monitors can be wired or wireless, and use ear buds for the performers. This eliminates any reasonable possibility of the monitor system causing feedback squeal, and absolutely avoids monitor wash. Some in-ear monitor systems allow each performer to have his/her own mix of other instruments and vocalists, and some of these systems permit each performer to control his/her own mix via a small control box on stage.

15.   Reverberation Enhancement

There are cases in which a performance or worship space may not provide enough reverberation to support the style of music to be used within the space. Examples are small rooms and/or rooms with low ceilings, cinemas that have been repurposed as general-use auditoria, and rooms having wooden ceilings in which cracks have opened up, changing the reflective surface into an absorptive one. There are also rooms that were incorrectly designed in the first place. There was a brief trend in church sanctuary design in the mid-20th century in which thin fiberglass panels were affixed to all side and rear walls. This resulted in an acoustically dull room which still had fairly high reverberation at lower frequencies. Likewise, many small rooms have been built using acoustically absorptive suspended ceilings, which not only kill the reverberation, but also decrease acoustical strength, and make even sound coverage difficult to achieve. The most straightforward way to enhance the reverberation in such rooms is to renovate them, replacing absorptive surfaces with reflective ones.

When renovation is not feasible for financial or other reasons, alternative ways of providing the desired reverberation have been devised. In 1941, Laurens Hammond patented his development of a spring unit that had originally been invented by Bell Telephone Laboratories to simulate the delay encountered in a long-distance telephone call. The "Hammond spring reverb" was successful in helping Hammond organs to sound more like pipe organs, despite the acoustically dead living rooms in which they were typically installed. Further developments of the "spring reverb" led to smaller units, and ultimately to units that appeared ubiquitously in guitar amplifiers and portable sound systems. The sound of these units, while sometimes considered pleasant, was unlikely to be confused with natural reverberation. In an attempt to create a more natural-sounding reverberation, the "plate reverb" was developed in the 1960's. While it was an improvement over the spring reverb, the plate reverb was not sufficiently natural-sounding to change the established practice of recording in a venue having the appropriate natural reverberation, when the project budget would support it. With the development of digital signal processing in the 1980's it became possible to create realistic-sounding reverberation electronically. As prices for these devices dropped, DSP reverb units replaced spring units in portable sound systems, and ultimately, in guitar amplifiers. Today, many mixers incorporate digital reverberation processors.

All of these artificial reverberation devices share two common problems. First, in a venue having real reverberation, the reverb sound is completely enveloping; whereas, when reverberation is added via the sound system, it comes from the speakers. This difference is noticeable. Second, sound-system-generated reverberation is not applicable to non-amplified instruments – most prominently, pipe organs, the instrument most dependent upon proper reverberation.

Beginning in the early 21st century, several manufacturers developed methods of enhancing the reverberation of a venue, using a multiplicity of microphones, amplifiers, and speakers, with a central DSP providing control. These systems are very expensive, but have achieved success in venues in which there were no other alternatives.

Some electroacoustical reverberation systems involve installing sufficient acoustical absorption in the room to reduce the RT60 to a fraction of a second.

Then they use an array of microphones to pick up the sound in the room, a DSP to create the reverberation through digital computation, and speakers to reproduce it. The microphones are then able to pick up the "reverberated" sound and recycle it through the DSP for another pass, and so on.

Other electroacoustical systems do not require deadening the room and creating artificial reverberation, but only enhance the natural reverberation of the venue. One example of such a system is the Yamaha Active Field Control (AFC) system. The difference in sound between the two approaches is that the enhancement systems incorporate the acoustical effects of the room's actual surface materials, retaining much of the room's sound. Commenting on a demonstration of the AFC system, one consultant summarized its effect by saying, "it's like doubling the height of the ceiling."

The cost of the electronic and electroacoustic (mics and speakers) elements in a reverb enhancement system is comparable to that of a very extensive sound system. The systems that replace, rather than enhance, the natural reverberation also involve the cost of installing large amounts of acoustically absorbing materials.

With either type of system, the user has the ability to vary the reverberation of the room within limits, to accommodate various types of music, and/or varying occupancies.

16. Variable Reverberation

There are many venues for performance and/or worship that regularly serve a variety of purposes. As discussed earlier in this book, the traditional way of handling such conflicts in acoustical requirements was to compromise. The preceding chapter introduced the idea of being able to vary the acoustics of a room as needed for a particular purpose: variable acoustics. Historically, there have been three ways of achieving variable acoustics: coupled spaces, reversible panels, and draperies.

Coupled Spaces

If two rooms are connected to each other through an opening that is small compared to the area of a wall, and a sound is made in room 1, the reverberation in room1 will begin to decay as it would it would if room 2 were not connected. Then at some point in time, the decay will change to what you would have in a single room having a volume equal to that of room 1 and room 2 added together. Thus, there is a "dual-slope decay" that provides a longer RT60 than room 1 would have by itself. Figure 16-1 shows the sound decay in a room with the sound source on a stage having full stage curtains, with the occupied auditorium coupled with a large reverberant room through an opening in a wall. The ragged-looking curve is the actual sound at the measurement location. The smoother curve above it is the averaged result from performing "Schroeder backward integration". The short, sharply-angled straight curve at left shows the RT60 of the stage area, based upon the decay rate established in the first 15 dB of the decay. The adjacent straight line shows the longer RT60 that includes the effects of the main auditorium. Note that if a third linear average were calculated, it would show an even longer RT60

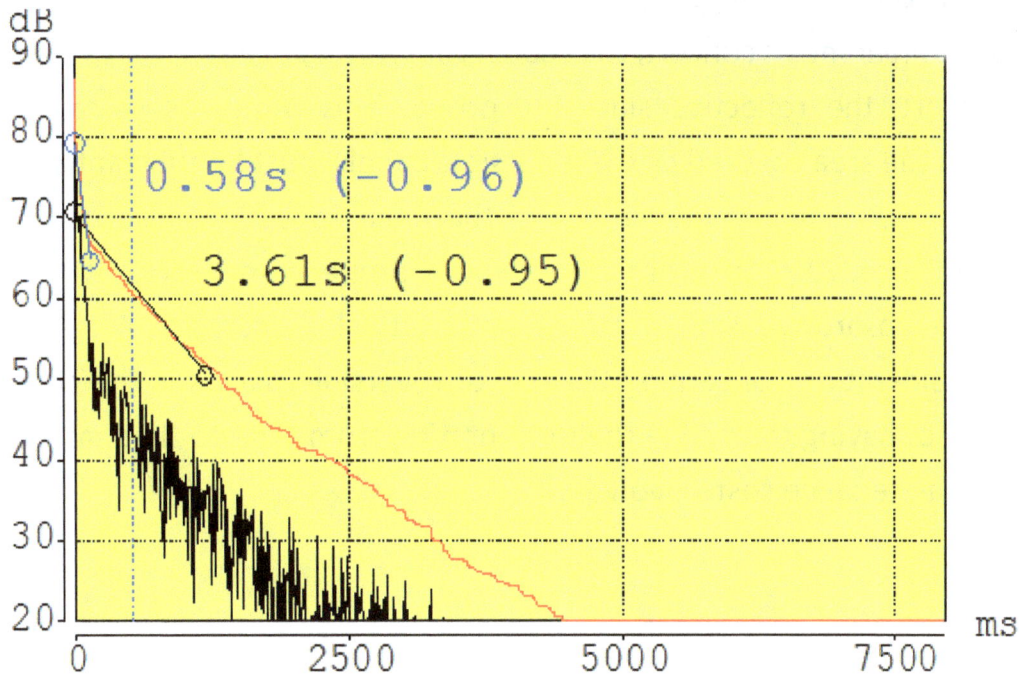

Figure 16-1: Multiple-Slope Decay

because of the concave curve of the decay line due to the coupled reverberant chamber. (This final composite RT60 would be about 4.9 seconds.) If operable doors are applied to the openings, the relative amounts of shorter and longer reverberation can be varied. A number of excellent concert halls have been built using coupled spaces. These are characterized by noticeably clearer sound than older venues having the same RT60, because of the rapid initial decay in a multiple-slope-decay room. In practice, a major obstacle to success with multiple-slope-decay design has been convincing the maintenance staff that all the empty volume of the reverberation chambers had to be kept empty, not used for storage!

Reversible Panels

While coupled spaces do present one option for variable acoustics in new buildings, if there is sufficient budget, there is another option that is less expensive and that can be applied during renovations. This method uses panels on some walls of the room that are acoustically reflective on one side and acoustically absorptive on the other. For a chamber music concert, the acoustically absorptive side is exposed. For a choral concert, the reflective side is exposed. This option is certainly not inexpensive, as the area covered by the reversible panels must be substantial, since it must have effects comparable to that of the main acoustical absorber in the room: the audience. Also, in order to achieve a proper variation of RT60 with frequency, the absorptive side must be rather thick – perhaps 4". Another important consideration concerning the use of reversible panels is the cost and inconvenience of having to change their orientation from time to time, as well as the possible maintenance costs involved.

Draperies

Covering the walls of a room with thick draperies seems at first glance to be worthy of serious consideration as a means of providing variable acoustics. Draperies may be less expensive than the other approaches. They are easy to pull, and permit an acoustical change to be made during an intermission. However, the acoustical effects of even heavy draperies are often overestimated. If folded to half their full length and hung a few inches from a wall, they are very effective absorbers of high frequencies. But no drapery can absorb low frequencies well. Thus, in the short-RT60 configuration (draperies extended), the room would sound overly warm, perhaps boomy. And certainly the room would sound dull because of excess absorption at high frequencies. Thus, as attractive as draperies may seem for providing variable acoustics, their performance does not support their apparent promise.

For both coupled spaces and reversible panels, changing the reverberation in the room is a time-consuming proposition, and neither method is capable of a change from a "rock room" to a "baroque concert hall". It is for this reason that electroacoustic methods are now considered the most practical method to achieve variable acoustics. It is now possible for a sound system operator to provide appropriate acoustics for almost any type of music at the flick of a switch, although the budgetary requirements are prodigious.

VI. Working with Your Room

17. Sound System Use

Introduction

This chapter is designed to provide general operating guidelines for using a sound system in a performance or worship space. For specific operating instructions to operate any piece of equipment, please refer to the owner's manual provide by the manufacturer.

System Power

Your sound system may be powered by a sequencer that is activated by a single switch

Figure 17-1: Power Sequencer/Conditioner (Photo courtesy of FSR, Inc.)

The figure above shows one model of power sequencer. The momentary-contact rocker switch is at left, designated **POWER**. To turn the system on or off, press and release this switch. If you hold the switch depressed, the system logic may become confused and power the system up, then immediately power it down again. If your system has remote power switches, they can be operated in the same way as the panel switch above.

The equipment should be connected so that the power amplifiers are last to be powered up, and first to be powered down. In this way, thumps caused by other equipment powering up or down are avoided. The power sequencer also provides surge protection to help avoid damage from lightning-induced power surges on the AC power mains.

System Layout

The block diagram on the following page shows the layout for a typical sound and video system including cameras for image magnification and video recording. Although your system may not include all the capabilities of this system (or it may have more capabilities), this diagram will illustrate the signal flow and interconnections for the sound and video. In a diagram of this type, signal always flows from left to right, unless arrowheads indicate otherwise. In the diagram, thin lines signify single connection cables, and thick ones indicate multiple cables, such as both right and left connections of a stereo pair.

The audio signal begins at microphones, either wired mics connected through floor boxes, suspended choir/baptistry mics, or wireless mic systems. Audio signals also come from CD/cassette players, as well as DVD/VCR's. Video signals may come from a computer, video camera, or DVD/VCR. Video signals are selected by a switcher-scaler, which also conditions the video for the projectors, recorders, and other downstream equipment. In a switcher-scaler, the audio follows the video, so if for example the DVD player is selected, its audio also goes to the audio output of the switcher-scaler. The audio+video sources are all selected by the switcher-

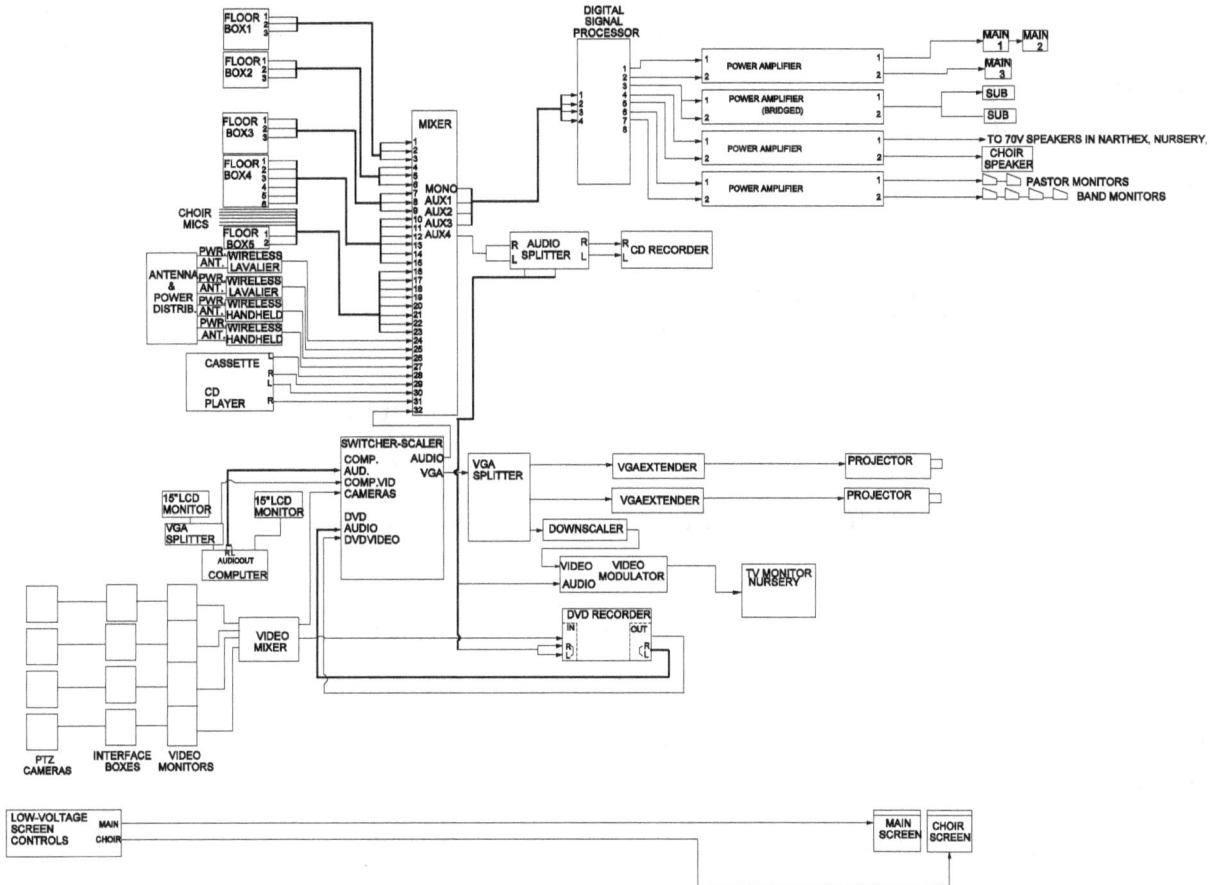

Figure 17-2: Typical Audio/Video System Layout

scaler, and their audio is then fed to a single mixer input pair (right + left), usually labeled **VIDEO**.

From the mixer, the audio goes to the digital signal processor (DSP) which provides tamperproof equalization, delay if needed, and speaker protection. This unit is adjusted at installation using special computer software. Under most circumstances, there is no need for the user to adjust the DSP.

From the DSP, the audio goes to the power amplifiers and then the speakers. The power amplifiers may be mounted in a separate rack from the other equipment.

125

Sometimes, powered speakers are used; these have the power amplifier(s) inside the speaker enclosure.

When multiple video cameras are used, the desired camera's signal must be selected by a video mixer, whose output feeds the switcher-scaler, or in some systems, a video recorder and/or remote TV monitors. Video signals from the switcher-scaler are split to feed projectors, recorders, etc.

Automix System with No User Controls

If your sound system has an automatic mixer without user controls, the block labeled **MIXER** in the diagram on the preceding page is in fact an automixer. Settings on this unit are made at installation, with a computer using special software. In general, any microphone, cassette/CD player, or other source connected to the mixer has its level turned up automatically when the automixer senses a signal. The resulting level of that signal is determined by the initial settings entered from the computer. If choir speakers or recording equipment are present, their levels are also controlled by the initial settings, although overall record level can be controlled by the level control on the CD or DVD recorder.

With a fully automatic system, the only user interaction needed is to power the system on and off as needed, load blank media into the recorders, set the recording levels, and make sure all microphones are properly connected.

If you experience a problem with operation of a fully automatic system, the troubleshooting grid below should help.

PROBLEM	STEPS TO TAKE
Pops, buzzes, hums, sizzle, no sound from a mic or other input device, or signal cuts in and out	1. Wiggle the cables one at a time where the connector attaches. If the problem changes, that cable should be replaced. 2. Disconnect source units (mics, wireless receivers, and CD/cassette players) one at a time. If the problem goes away, the cable or source unit is defective. Try replacing the cable. If it does not fix the problem, the source unit is defective. 3. If you have determined that a wireless microphone system is at fault, try changing the channel on both transmitter and receiver per the instructions in the manufacturer's manual.
No sound at all	1. Verify that power lights on all equipment are illuminated. 2. Verify that level controls on power amplifiers have not been turned down. 3. See whether any **THERMAL** or **FAULT** indicators on the power amplifiers are lighted. If so, call a service technician.
Feedback squeal	1. Verify that microphones are not being used directly in front of speakers. 2. Verify that level controls on power amplifiers have not been turned up too high.

If you are not able to correct the problem using the above steps, call a service technician or an audio consultant.

Automix System with User Controls

Automatic-mixing systems with user controls allow some adjustment of levels for individual mics and other sources, but usually do not allow the user to turn sources on too loud or completely off.

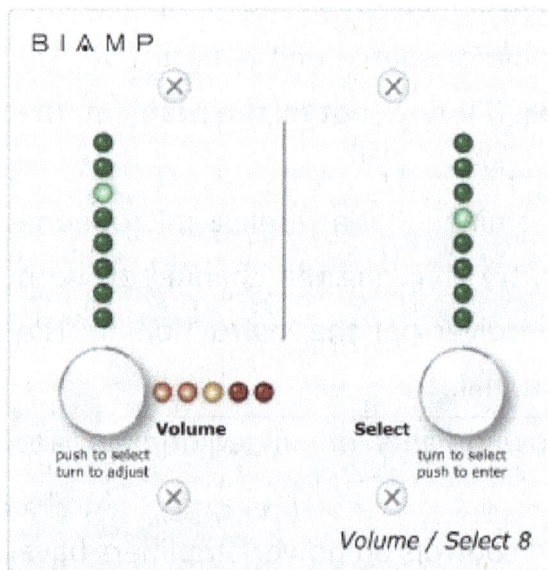

Figure 17-3: Biamp User Control (courtesy of Biamp Commercial Audio Systems, Inc.)

With the Biamp control shown at left, you push the **Volume** knob until the green light designated for the desired mic or other source is illuminated. There will be labels beside the green lights. Then you can adjust the level of that input, with the level being indicated by the row of red lights beside the knob. The **Select** knob can be rotated to select a function, such as a preset change. Then the function is executed by pushing the knob.

Some systems have custom-built user controls that may operate differently than the one shown in Figure 17-3. And some automixer have built-in controls, as shown in Figure 17-4.

Figure 17-4: Electronics Digital Matrix Processor (Auto mixing DSP) Model DM84 (Photo Courtesy of Lectrosonics, Inc.)

The troubleshooting grid shown on the previous page should help if you have trouble with an audio mixing system with user controls.

Manual Mix Systems

Any audio mixer, whether digital or analog, performs certain functions. The drawing in Figure 17-5, below, illustrates some of these functions.

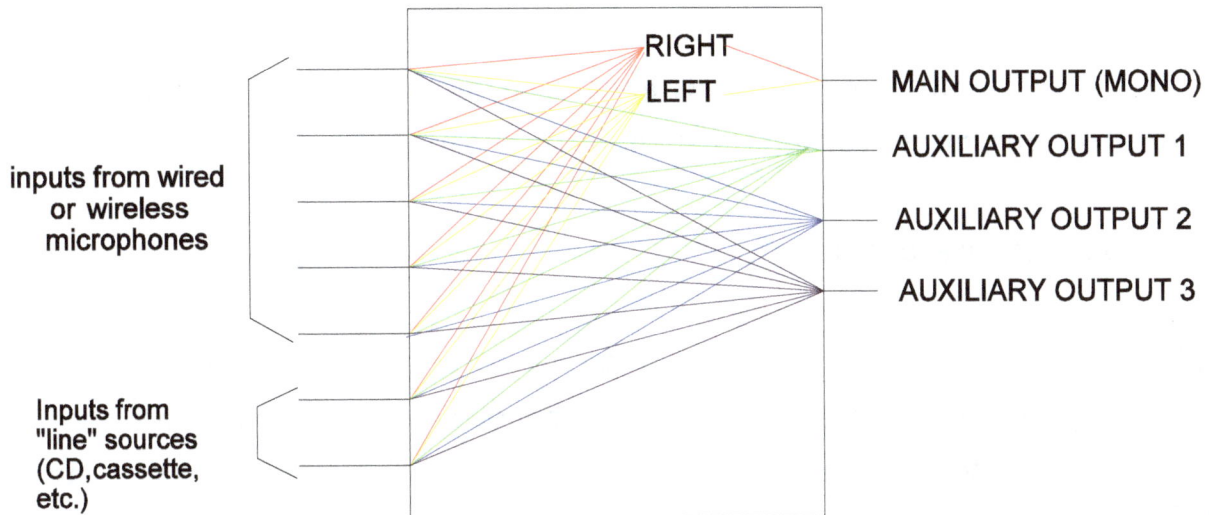

Figure 17-5: Basic Mixer Functions

Note that the device we call a "mixer" is really several mixers in one box: this one mixes from seven sources to any of four outputs or "mix buses". The main output feeds the main speakers. The auxiliary outputs feed choir speakers, floor monitor speakers, recorders, and hearing assistance devices.

In addition to the actual mixing, a mixer allows tailoring of the sound of each source via equalization ("tone controls") and sometimes compression and special effects. It is convenient to discuss mixer operation in terms of individual channel sections ("channel strips") and the main sections. Since these controls operate somewhat differently for analog and digital mixers, each type of mixer will be discussed separately. In order to avoid necessitating frequent paging back and forth when a reader is using this chapter, I will include some information in both the analog and the digital mixer sections.

No matter what type of mixer is used, chances of feedback squeal are minimized if only the mics being used at a given time are "hot." Keep unused mics muted (not "**ON**") or their sliders down.

Digital Mixers

Note: The information below is general in nature. For detailed operating instructions, please consult the manufacturer's manual.

Basic channel operation

Many digital mixers use "layers" to allow control of many channels in a physically small console. Thus a 48-channel digital mixer may have only 24 channel strips, arranged on two layers. In order to control a single channel, you must first select the correct layer. This is done by pressing a button that assigns the controls for a group of channels or a mix bus to the controls on the channel strips or on a "selected channel" control cluster.

Once the correct layer has been selected, a channel must be selected before it can be controlled by the knobs. Figure 17-6 illustrates the channel strip of a popular digital mixer. The **SEL** button at the top selects the channel. The **CUE** button feeds the signal from that channel into the headphones, and allows the operator to monitor the level of the signal on the LED meter. The **ON** button must be pressed to turn the channel on, or no signal from that channel will reach any of the mix buses. When this button is pressed, the green LED inside the button is illuminated. Many mixers have a **MUTE** button instead of an **ON** button. When this button is pressed, a red LED is illuminated, indicating that the channel is muted and there will be no output until it is unmuted. The sliding control at the bottom is called a "fader," and controls the level of signal from the channel into the main

Figure 17-6: Channel Strip for Digital Mixer

mix. Additional controls for the selected channel are accessible through the controls in the "selected channel" cluster.

Almost all mixers have a way to set the input sensitivity for each channel. Variously referred to as **TRIM**, **GAIN**, **HA GAIN** (for "head amplifier gain") or **ATTEN** (for "attenuation"), on a digital mixer this may be a hardware control located above the channel strip, or a software-activated control on the LCD screen. If this control is set at minimum, there will be no signal from that channel. It should be adjusted so that the fader produces a normal output level for the channel when the fader is set around 0 to -10 dB.

Many microphones require DC power for their operation. This power is provided by the mixer, and is called "phantom power." Phantom power can be switched on or off with either a single switch for the whole mixer; several switches, each controlling a group of channels; or a switch on each channel. On digital mixers, phantom power is usually controlled through a hardware switch or a software switch on each channel. If phantom power is turned off, many microphones will not produce a signal to the mixer.

Basic Main Bus Operation

The main channel on a digital mixer is much like the input channels: it has a fader marked something such as **MAIN, MASTER,** or **STEREO,** plus a **SEL** button and an **ON** or **MUTE** button. The **ON** or **MUTE** button activates or mutes the main output of the mixer.

Basic Monitor Bus Operation

There are two functions denoted by the word "monitor" in relation to mixer operation. In a mixer designed for recording, "monitor" usually means the control room monitor speakers. In a mixer designed for live sound, "monitor" usually means a "performance monitor," which is a speaker used to allow performers or worship leaders to hear themselves. Control room monitor sections on mixers are seldom of much use in live sound, so we will limit our discussion to performance monitors. Generally, these are auxiliary mix channels which are fed by **AUX** software controls of the individual channels. On some digital mixers, these controls are designated **MIX/MATRIX** rather than **AUX**. The levels of the various channel signals appearing in the monitor mix are set in the individual channel controls. Usually, there is also a master level control for each **AUX**.

Each **AUX** send can be set so that it is affected by the fader setting for the channel ("Post-Fader") or so that it is independent of the fader setting ("Pre-Fader"). Stage monitors are usually set to **PRE** so that the monitor mix is independent of the main mix.

AUX or **MIX/MATRIX** functions are also used to feed recorders and hearing assistance equipment.

Troubleshooting

The troubleshooting grid below should help if you have trouble with a sound system incorporating a digital mixer.

PROBLEM	STEPS TO TAKE
No sound from any mic or other input device	1. Make sure the main output is not muted: **ON** switch is activated, or **MUTE** switch is not engaged. 2. Verify that power lights on all equipment are illuminated.

	3. Verify that level controls on power amplifiers have not been turned down.
	4. See whether any **THERMAL** or **FAULT** indicators on the power amplifiers are lighted. If so, call a service technician.
No sound from one wired mic or other input device	1. Make sure the channel is not muted: **ON** switch should be activated, or **MUTE** switch should not be engaged. Make sure phantom power on the channel is on.
	2. Disconnect the mic or other input device and plug it into a different channel. If it works in the new channel, call a service technician. If it does not work in the new channel, try replacing the cable. If that does not fix it, the source unit is faulty.
Pops, buzzes, hums, sizzle, or signal cuts in and out	1. Mute the mixer channels one at a time. If you are able to stop the disturbing noise, proceed to the next step. Otherwise, call a service technician.
	2. Wiggle the cables one at a time where the connector attaches. If the problem changes, that cable should be replaced.
	3. Disconnect source units (mics, wireless receivers, CD/cassette players) one at a time. If the problem goes away, the cable or source unit is defective. Try replacing the cable. If it does not fix the problem, the source unit is defective.
Feedback squeal	1. Verify that microphones are not being used directly in front of speakers.
	2. Verify that level controls on power amplifiers have not been turned up.
	3. Verify that channel equalization controls are not set to boost any frequencies by more than a few dB.

Analog Mixers

Note: The information below is general in nature. For detailed operating instructions, please consult the manufacturer's manual.

Channel strips

Figure 17-7 illustrates the channel strip of a popular analog mixer. The sliding control at the bottom is called a "fader," and controls the level of signal from the channel into the main mix. Beside the fader are three sets of "assign" buttons: **L-R, 1-2,** and **3-4**. These buttons determine where the output from the channel goes: to left and right, to groups 1 and 2, or to groups 3 and 4. Groups are also called "sub mixes," which is a descriptive name for them. Groups will be discussed further in the section on the mixer's Main section. Above the fader is a small LED level meter for the channel. When the mixer is in use, the levels should rarely go into the yellow region, and if they illuminate the red lights, the **TRIM** or **GAIN** control for the channel should be turned down.

Above the fader is the **PFL** button, called the **CUE** button on some mixers. **PFL** stands for "pre-fader listen"; that is, pressing this button allows you to listen to the signal of that channel on the headphones, regardless of the fader position. Next is the **MUTE** button. When this button is pressed, a red LED is illuminated, indicating that the channel is muted and there will be no output from the channel until it is unmuted. (Instead of a **MUTE** button, some mixers use an **ON** button which must be pressed to turn the channel on). The control above the **MUTE** button is the **PAN** control. This knob adjusts how much of the channel's signal goes to the left, to group 1, or to group 3, *vs.* how much goes to the right, to group 2, or group 4.

Figure 17-7: Channel Strip for Analog Mixer

The next six knobs control how much of the channel's signal goes to each auxiliary mix, as discussed in the section on Monitor mixes. The two pushbuttons associated with the **AUX** knobs will also be discussed in that section.

The next six knobs are for equalization ("tone control"). The bottom one of the group, **LF**, is the bass control. The next four comprise the Low Midrange and High Midrange controls. These controls are paired: there is a blue-topped knob that controls boost or cut, and a green-topped knob that controls the frequency where the boost or cut takes effect. Just above the equalization controls is an **HPF** pushbutton. This "high-pass filter" removes low-frequency rumble such as noise from bumping mic stands.

Almost all mixers have a way to set the input sensitivity for each channel, variously referred to as **TRIM**, **GAIN**, **HA GAIN** (for "head amplifier gain") or **ATTEN**. If this control is off, there will be no signal from that channel. It should be adjusted so that the fader produces a normal output level for the channel when the fader is set around 0 to -10 dB. Associated with the **GAIN** control of this mixer is a **LINE** pushbutton which adapts the channel to receive line-level inputs, as from a CD/cassette player.

The **POLARITY** pushbutton is important when two mics are located close together, and determines whether the signals from the mics add or cancel each other. Normally, this pushbutton should be in the "out" (non-activated) position.

Many microphones require DC power for their operation. This power is provided by the mixer, and is called "phantom power." Phantom power can be switched on or off with either a single switch for the whole mixer; several switches, each controlling a group of channels; or a switch on each channel. On this mixer, phantom power is controlled through the **+48V** switch on each channel. If phantom power is turned off, many microphones will not produce a signal to the mixer.

Main section

Figure 7-8 shows the main section controls of a popular analog mixer. There are faders for the left, right, and mono (**L+R**) outputs. The **MUTE** button mutes the main output of the mixer. Some mixers have an **ON** Button instead, which must be pressed to activate the main output. The **AFL** (after fader listen) button sends the signal from the selected output to the headphones. A small LED level meter shows the signal level, which may also be seen on the full-sized level meter typically located in the upper right quadrant of the mixer.

136

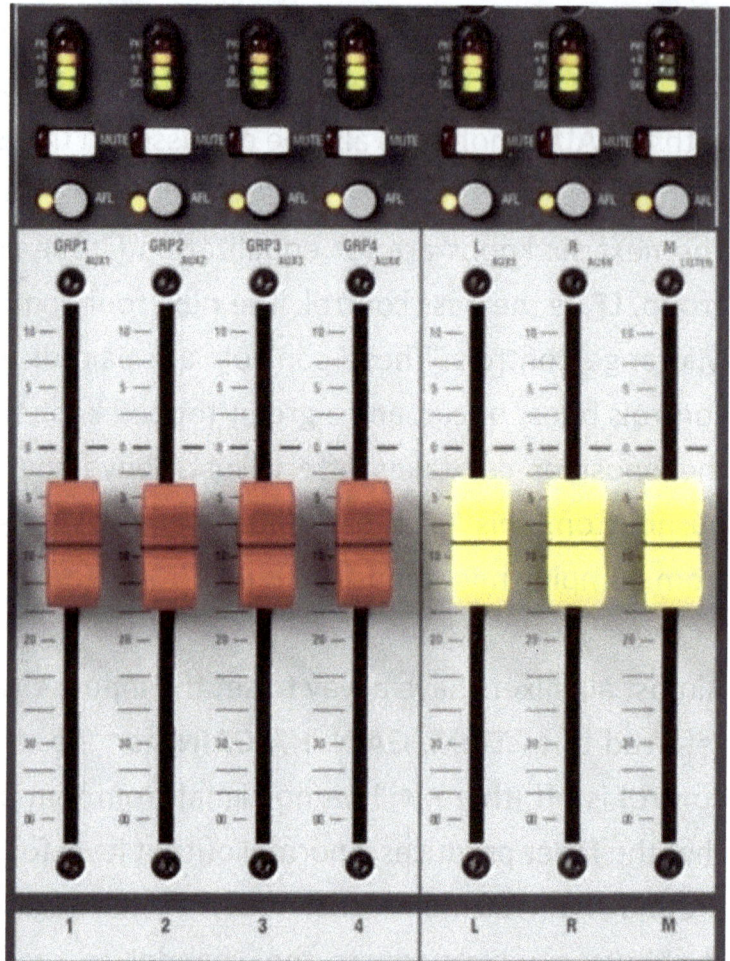

Figure17-8: Main Section Controls

Sub mixes

As mentioned earlier, many mixers provide additional "groups" or "submix buses" that can be used to simplify the operator's task. For example, let us assume that mics 2-5 are being used by a singing quartet: 2–soprano, 3-alto, 4-tenor, and 5-bass. During rehearsal, a good mix is established so that the singers are properly balanced. Later during the performance or worship service, it is convenient to be able to raise or lower the entire quartet without disturbing the balance among the singers. This can be done by assigning channels 2-5 to a single group; let's say, group 1. This is done by pressing the **1-2** buttons on channel strips 2-5, and making sure the **L-R** buttons are released. (As a general rule, it doesn't make sense to assign a given channel to more than one mix bus pair.) Also, the **PAN** controls on channels 2-5 should be fully counter-clockwise so that the signal goes to group 1 and not to group 2. Having done this, you can now control the level of the entire quartet by the **GRP1** fader. Groups are also handy for mixing drum sets and small instrumental ensembles.

Monitor mixes

There are two functions denoted by the word "monitor" in relation to mixer operation. In a mixer designed for recording, "monitor" usually means the control room monitor speakers. In a mixer designed for live sound, "monitor" usually means a "performance monitor," which is a speaker used to allow performers or worship leaders to hear themselves. Control room monitor sections on mixers are seldom of much use in live sound, so we will limit our discussion to performance monitors. Generally, these are auxiliary mix channels which are fed by **AUX** software controls of the individual channels. On some digital mixers, these controls are designated **MIX/MATRIX** rather than **AUX**. The levels of the various channel signals appearing in the monitor mix are set in the individual channel controls. Usually, there is also a master level control for each **AUX**.

Each **AUX** send can be set so that it is affected by the fader setting for the channel ("Post-Fader") or so that it is independent of the fader setting ("Pre-Fader). On the channel strip shown in Figure 17-7, there is a pushbutton to set **AUX 1** and **2** as **PRE** or **POST**, and another pushbutton for **AUX 3** through **6.** Stage monitors are usually set to **PRE** so that the monitor mix is independent of the main mix.

AUX or **MIX/MATRIX** functions are also used to feed recorders and hearing assistance equipment.

Troubleshooting

The troubleshooting grid shown after the digital mixer section (page 132) should help if you have trouble with a sound system incorporating an analog mixer.

Combined Automatic/Manual System

A combined automatic/manual system is useful for providing both the convenience of an automatic sound system for simple events, and the versatility of a manual system for more complex events. In order to make use of the superior microphone preamplifiers contained in a manual mixer, the input sources are connected directly to the manual mixer. Channel direct outputs are used to feed the channel signals to the automatic mixer. Different presets are used in the digital signal processor (DSP) to send the outputs of the proper mixer to the power amplifiers for the automatic and manual modes. Except for the need for signal to pass through both mixers in the automatic mode, the section on Automix Systems with No User Controls and the section on Manual Mix Systems apply to these combined systems. The only problem specific to a combined system is that if the main or channel outputs of the manual mixer are muted, or if the **TRIM**, **GAIN**, or **HF GAIN** controls are set at minimum, there will be no output from either the manual or the automatic system.

Common Equipment

Wireless Microphone Systems

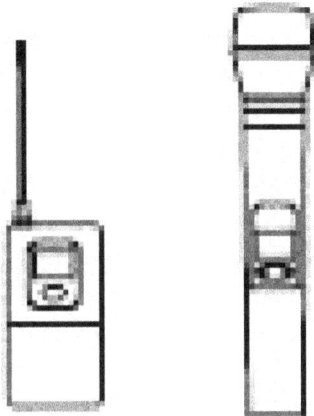

Figure17- 9: Wireless Receiver

A wireless microphone system is made up of a transmitter such as those shown in Figure 17-9, and a receiver such as the one in Figure 17-10. The transmitter and receiver must be set to operate on the same frequency, and that frequency must not be the same as that of any other wireless systems or other radio or TV transmitters nearby. Your wireless frequencies were set up when your system was installed, and probably will not need to be changed. However, instructions on setting wireless system frequencies are included in the manufacturer's manual.

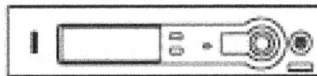

Figure 17-10: Wireless Transmitters

Most wireless systems today use AA batteries, although a few still use 9V ones. Alkaline batteries are preferred for best performance and longest life. The manufacturer's manual will give estimated battery life for your transmitters. Many users are curious about using rechargeable batteries in order to save money. Rechargeable batteries that would fit wireless transmitters come in two technologies: Nickel-Cadmium (NiCd), and Nickel-Metal Hydride (NiMH). NiCd batteries are plagued with "memory" which causes them to lose their ability to take a full charge unless they are always fully discharged before being recharged. NiMH batteries do not have this defect. However, neither rechargeable battery can provide the full 1.5V supplied by an alkaline battery; rechargeables put out more like 1.2 – 1.3V. This lower voltage reduces the effective radio-frequency power output of the transmitter, reducing its range, and increasing likelihood of noise due to weak signal.

The chart below provides basic troubleshooting information for wireless microphone systems.

No sound from one wireless mic	1. Make sure the receiver for the wireless mic is **ON**. 2. Make sure the channel is not muted: **ON** switch is activated, or **MUTE** switch is not engaged. 3. Make sure the transmitter battery is good. Remember that there is such a thing as a new, bad battery. 4. Make sure that the wireless transmitter and receiver are set to the same channel, and that other wireless systems are not set to that same channel. Refer to the manufacturer's manual for instructions on setting channels. 5. Disconnect the wireless receiver and plug it into a different channel. If it works in the new channel, call a service technician.
Scratchy noise or signal cuts in and out (belt pack transmitter)	While the mic is on and the system is live, gently wiggle the cable between the lavaliere or earset microphone and the transmitter. Wiggle it first near the mic element, and then near the connector. If wiggling the cable makes a difference in the performance, the cable will need to be replaced or repaired or the microphone/cable assembly will need to be replaced. Call a service technician.

CD and DVD Recorders -- Audio

Mixing the recording

Mixing a recording is much like monitor mixing, in that an **AUX** send from each channel goes to an auxiliary bus which feeds the recorder. The best way to get the proper mix is to set the levels while using a headset connected to the recorder.

Alternatively, the headset can be connected to the mixer, with the headset source control set to the auxiliary channel used for recording. It is best to listen to a completed recording later in order to refine the mix for future recordings in the same venue, if necessary.

The record level should be set while you are looking at the level meter on the recorder. During the loudest portions of the program, the meter should never, or very rarely, bump into the red region, or the recorded sound will be very distorted. At the same time, it is important to have the record level high enough so that the listener does not have to turn his/her sound system up too far in order to hear well, so the loudest peaks should be near the red region.

Recording to CD or DVD

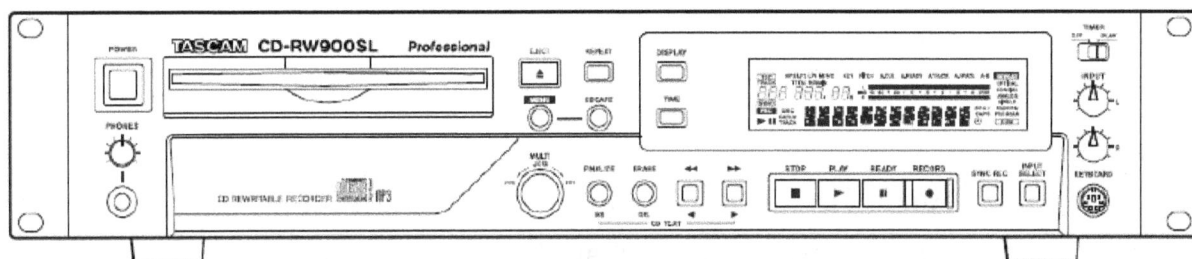

Figure17- 11: CD Recorder (Photo Courtesy of Tascam, Inc.)

Figure 17-11 shows the front panel of a professional CD or DVD recorder. With most CD or DVD recorders, there is an **INPUT SOURCE** switch which must be in the **ANALOG** position. You start recording by pressing the **RECORD** button. Then you can view the signal level on the recorder's level meter, and the recorder senses the type of disk and initializes itself. Recording actually begins when you press **PLAY**. If you want to start a new track to make it easier for listeners to find specific portions of the program or worship service, you can press **RECORD** just before the new section of the program begins. Most CD or DVD recorders can be set up to automatically start a new track whenever a silence of a certain duration is detected. This feature requires careful setup so as not to respond to pauses between sentences, etc. The manufacturer's manual contains details on setting up this feature.

Once recording is completed, the CD or DVD must be finalized before it can be played. This involves inserting table of contents and other data that the CD or DVD player can read. Finalizing is initiated by pressing the **FINALIZE** button. The recorder may ask for confirmation once or twice; you can confirm by pressing **ENTER** on the remote control, or, on the machine shown in Figure 17-11, pressing the **MULTI-JOG** control. The recorder display will tell you when finalizing is complete.

Troubleshooting

CD's won't record	Make sure the **INPUT SOURCE** switch is in the **ANALOG** position.
CD's or DVD's seem to record but won't play back	Look in manufacturer's manual for instructions on finalizing disks after recording.
CD's or DVD's that have been recorded seem weak or distorted on playback	For weak CD/DVD sound, turn up record level on recorder. If playback is distorted, turn record level down. During recording, the level LED's on the recorder should bounce as close to the red as possible without ever illuminating the red lights.

Hearing Assistance

Hearing assistance systems are usually set up upon installation, and require no user intervention. User receivers have volume controls that may be adjusted to the user's preference.

Other than battery replacement, hearing assistance systems require little maintenance. Occasionally the ear bud or ear speaker cable is damaged by excessive flexing, but even that problem is rare.

The signal level for the hearing assistance system was set during system installation. If some of the sources do not seem loud enough, you may need to adjust the mix. This is done in the same way as adjusting the mix for recording. (See the earlier section on mixing the recording.) In fact, many systems use the same signal for both recording and hearing assistance.

Effects

How Connected

Most venues for traditional concert music, or for traditional or blended worship have enough natural reverberation, and little need for extensive production effects during a worship service, so that no effects units are needed. Churches using contemporary worship, however, often need to use reverberation, compression, or other effects. There are three common methods of applying effects units.

The simplest way to use effects is to take advantage of the effects unit built into the mixer, if your mixer has one. Typically, there are one or two auxiliary mixes that are specifically designated for effects. For each channel on which the effect is desired, the effects send control is turned up enough to provide the desired level of the effect. The overall amount of the effect (typically, reverberation) from all sources is controlled by an effects return knob. Digital mixers usually provide abundant effects. If you have a digital mixer, please consult the manufacturer's manual for guidance in setting up effects.

For mixers that do not include effects, one **AUX** channel can be dedicated to effects. Then the outboard effects unit is fed from the corresponding **AUX SEND** connector on the back of the mixer, and the output of the effects unit is fed into one of the mixer's line inputs. The operation of this approach is just like the one described above, except that the overall effects level is controlled by a channel fader rather than an effects return knob.

A third way to apply effects is useful when an effect is needed on only one channel. In that case, the **INSERT** jack on the back of the mixer can be used with a special insert cable to feed the signal from the desired channel into and out of the effects unit. The effects level is controlled by knobs on the effects unit itself.

Troubleshooting

No effect heard	1. Check settings of effect or aux control on channel strip.
	2. If effects unit is outboard (not part of the mixer), wiggle the cables. If the problem changes, replace the cable.
	3. If the effects unit is built into the mixer, check the settings of the effects section on the mixer, using the manufacturer's manual as a guide.
Hum, buzz, or noise when effect is used	1. Check the cables as above.
	2. Unplug the effects unit from the AC power. If this stops the buzz, disconnect the input to the effects unit from the mixer and replug the effects unit to AC. If the noise comes back, replace the effects unit. If not, replace the input cable.

Special Functions

In recent years, the use of pre-recorded accompaniment tracks instead of live musicians has become increasingly common. Often these tracks are provided in a left-right format, but it is not true stereo. This left-right format typically has only the accompaniment on one channel – e. g., the right channel; and a recorded vocalist or choir on the other channel. It is important for the sound operator to make sure that only the desired channel is fed through the mixer. On mixers having stereo inputs, this can be done via the **BALANCE** control, which in our example would be set fully to **R** if only the accompaniment is desired. The signal from that channel is then fed (usually through an **AUX** channel) to stage monitors arranged so that the vocalist(s) can hear them, and the level is adjusted to suit the performer(s). Then if further accompaniment sound is needed in the audience/congregation area, the main fader on the cassette/CD channel can be brought up appropriately. It is not uncommon for the performers to need the accompaniment to be loud enough on the monitors that the main fader does not have to be advanced at all, except perhaps to provide a more natural sound than that which comes from monitor wash.

Electrical/Electronic Instruments

Figure 17-12: Rapco db-1 Direct Box

The traditional way to amplify instruments such as electric guitar was to use a guitar amplifier with a microphone placed within about a foot of the front, and pointed at the speaker. In this way, the amplifier's characteristic sound was heard, only louder. Today, many electric guitarists and bassists prefer to use a small electronic unit that emulates the sound of their favorite amplifier, and can be plugged directly into the sound system. Essentially this same approach is used for electronic keyboards. In both cases, the instrument's signal has a higher voltage than a microphone, and it is almost always "unbalanced", meaning that it is more susceptible to hum and buzz than are the other sound system components. To avoid the voltage level incompatibilities and reduce the hum and buzz, **direct boxes** (Figure 17-12) are usually used as an interface between an electrical or electronic instrument and the mixer of a sound system. The instrument's cable is plugged into the input of the direct box, and the output of the direct box is connected to one of the mixer's microphone inputs via a microphone cable.

It is very seldom appropriate to connect a self-contained electronic organ (one having its own amplifier and speakers) to a sound system. A sound system is designed to provide accurate reinforcement of the sound it is reinforcing; whereas, an organ amplifier/speaker system are designed to provide specific "voicing" that the organ manufacturer considered appropriate for that particular model. Thus, the sound of an organ heard through a sound system seldom does justice to the instrument. Attempts have been made to use organ amplifiers and speakers for general purpose sound systems, with even more disastrous results.

18. Using Portable Sound Systems

146 For outdoor performances or worship services, or for such events that are held in rooms lacking an adequate sound system, portable systems must be used. Despite the great variability among models of portable sound systems, such systems can be generally classed as one of the following: podium systems, general-purpose systems, and touring concert systems. These are illustrated in Figures 18-1 through 18-3.

Figure 18-1: AmpliVox Presidential Plus Podium Sound System (Photo courtesy of AmpliVox Sound Systems, Northbrook, IL, www.ampli.com.

Podium sound systems contain an amplifier and speakers inside a podium, atop which is a microphone mounted. These systems have the advantage of being extremely easy to set up. However, the proximity of the microphone to the speakers creates difficulties in achieving adequate sound level without feedback squeal. Also, mechanical vibrations from the speaker can be conducted directly to the microphone, providing another mechanism for producing feedback squeal. In order to reduce these problems, manufacturers of such systems often deliberately limit the low-frequency response, making the sound somewhat harsh and unnatural. For small audiences in fairly small rooms, the better podium sound systems can perform adequately, provided they are used only for speech reinforcement.

147

Figure 18-2: Yamaha Stagepas500 Portable Sound System (Photo Courtesy of Yamaha Corporation)

The general-purpose portable sound system shown in figure 18-2 consists of a small mic-stand-mountable mixer/amplifier and two stand-mountable speakers. Many manufacturers produce variants of this type of portable sound system. Some utilize powered speakers (speakers incorporating power amplifiers); some include the mixer in a powered speaker; some are much larger and include desk-type powered mixers. These systems can perform well as long as the audience area is not too large – "too large" depends upon the characteristics of the specific system. As with all sound systems, these cannot cover as large an audience outdoors as they can indoors, because of the greater rate of sound level decrease with distance outdoors, due to absence of room gain outside, and often because of greater ambient noise outdoors.

Figure 18-3: SE Systems Stage Van Touring Sound System (Photo Courtesy of
Terry Richardson, SE Systems, Inc.)

Touring concert sound systems are usually owned by production companies, and can be rented, along with setup crew and operators, to cover audiences numbering in the thousands. Their operation is much like a sophisticated installed system.

Setting up a portable sound system begins with choice of the speaker locations, which should be where the every person in the audience can clearly see at least one speaker. Insofar as possible, the speakers should be located so as to make the speaker-to-listener distance as nearly equal for all listeners as possible. In this respect, altitude helps: the higher the speakers are positioned, the easier it is to keep the sound loud enough for rear listeners without blasting those in front.

In planning the speaker location, remember that the speakers should not play into the microphones, or you will likely have problems with feedback squeal. If the performers require stage monitor speakers, arrange to have dedicated monitor speakers for this purpose; don't try to make the main speakers do double duty.

The second step is to select the control location so that the sound operator can hear what the audience hears. Having the operator not be seen is far less important.

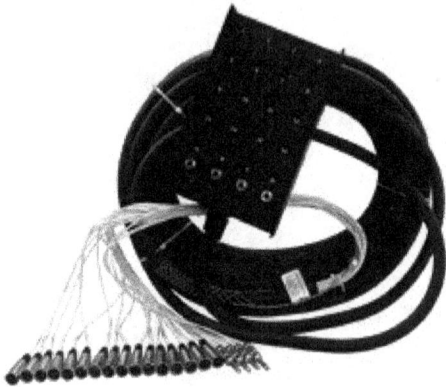

Connections from the microphones and electrical instruments on stage to the mixer can be made using ordinary microphone cables, but it makes for a neater appearance if a single multi-conductor "snake" cable is used. In this case, mics and instruments are plugged into a "stage box" and connectors at the other end of the snake are plugged into the mixer. "Power snakes" are also available; these include cables that connect the amplifiers to the speakers.

Figure 18-4: Rapco S-Series Snake

150

The main difference between operation of a portable system and an installed system is that portable systems are more likely to use powered mixers, which may offer less flexibility than most installed mixers. But basic operating principles are the same. Troubleshooting, however, is a bit different, since the individual microphone and instrument cables, as well as the snake, are exposed to more flexing than their counterparts in an installed system; therefore, in troubleshooting a portable sound system, first suspect the cables.

19. The Musician/Vocalist Environment
Using Monitor Speakers

Other than using the appropriate monitor speakers, and positioning them so that the performer can hear as well as possible and so that they do not feed back into the microphones, the other primary consideration relating to the use of monitor speakers is the mix of signals fed to the monitor, and the level at which the monitor is operated. The purpose of a stage monitor system is to allow the performers to hear each other – and to some extent, themselves. The system performs this function best when as few instruments and vocalists as possible are included in the monitor mix. If acoustic drums are used, they probably do not need to be included in monitor mixes, even if they are miked for the audience. If electronic drums are used, at least the kick drum and snare are probably needed in all the monitor mixes, to help the performers stay synchronized. It is also helpful for all performers to hear the bass (electric, string, or keyboard) well, as it forms the basis for tuning. Depending upon the size/presence of the amplifier for the bass, this may or may not mean that the bass must be in the monitors.

All vocalists should be able to hear each other in the monitors, except when the vocal ensemble is a choir. In this case, they need to be able to hear each other acoustically, since miking a choir and then feeding the mic signal back into the choir monitor speakers is an almost sure recipe for feedback squeal.

Beyond these general rules, the choice of instruments in the monitor mix depends heavily upon instrumentation, musical arrangements, stage layout of the performers, and personal taste. Where floor ("wedge") monitors are used, keeping the number of instruments fed to each monitor to a minimum will decrease problems with monitor wash.

No matter whether monitor speakers or in-ear monitors are used, it is beneficial for the monitor levels to be kept as low as possible. The difficulty in doing this stems from the excitement produced by loud music, and performers can become addicted to this excitement. Some public speakers even request that monitors be provided so that they can hear themselves; often the reason is to provide an emotional surge that these public speakers feel they need to experience in order to be most effective. However, loud monitor speakers inevitably produce monitor wash. Monitor wash, coming off the back of the monitors, is by nature muddy sounding, making for a poor listening experience for listeners seated near the front of the room. Many times, in an effort to overpower the monitor wash, system operators will increase the main sound system levels, eventually leading to sound levels that can be damaging to the listeners' hearing.

Likewise, users of in-ear monitors who insist on high monitor levels can cause damage to their own hearing. This is particularly insidious, since in a loud room, we sense the loudness through other parts of our bodies, not only the ears. Without this extra sensory input, we do not as readily notice how loud a sound is, so with earphones or ear buds, we can be exposing our ears to very dangerous levels without even noticing it. By the time ear pain begins, permanent damage has already occurred.

Rehearsal in Empty Rooms

152

Earlier in this book, the differences between the sound of an empty room and an occupied room have been discussed: the RT60 of an empty room will be longer than that of an occupied room, especially at middle and high frequencies. Instrumentalists and vocalists who must rehearse in empty performance spaces are aware of the difference in sound. In one way, this difference is helpful, since it encourages performers to pay close attention to articulation. However, rehearsing in a too-reverberant space can also cause a performer to become accustomed to an excessively staccato style in an effort to make the music sound good in spite of the reverberation, with the result that the performance can sound choppy when a full audience is present. Of all instrumentalists, probably organists are most affected by this phenomenon. The only solution to this dilemma is that in rooms using electronic reverberation enhancement, the enhancement system should be used during rehearsals as well as during performances.

Preferably, the enhancement system should be set up with presets for "empty", half-occupancy", and "fully occupied" conditions. The appropriate preset can then be chosen for each rehearsal or performance.

20. Correcting Poor Acoustics in Existing Rooms and Maintaining Good Acoustics Through a Renovation

General Renovation Issues

When a venue for performance or worship is renovated, two questions should be asked at the outset:

1. Is the program material to be performed in the renovated facility the same as in the old facility?
2. If so, does the room as it is properly support that program material from an acoustical standpoint?

If the program material will change, the renovation goals should reflect any needed changes in acoustics to support the new purpose of the room. Examples of this principle being ignored abound. During the mid-20th century, many old cinemas were converted into civic centers, theaters, and performance halls. In most cases, the abundance of soft, acoustically absorptive materials that had served successfully in the cinemas were kept in place, resulting in rooms with insufficient reverberation and acoustical strength for many functions. These materials included heavy velour stage curtains, heavily upholstered seats, thick floor carpets, and even carpeted walls. After a successful community fund drive had been conducted to finance the renovation of an old cinema, going back to donors for additional funds to correct the acoustics (as an afterthought) would have been politically very difficult.

If the program material will not change, but the room needs acoustical improvement, a renovation is the ideal time to address this need. Renovations often include new floorcovering, wall and/or ceiling repairs, and painting. Improper materials choices in any of these areas can adversely affect the acoustics. Often existing acoustical panels are found to appear dirty or dingy, and there is a desire to paint them. Doing so will destroy the acoustical properties of most absorptive panels, but the panels can be replaced with new ones, creating an opportunity to engage a consultant to review the material, thickness, and placement of the panels. Common wall finishes such as gypsum boards ("sheet rock") and thin wood paneling can provide a lot of acoustical absorption at low frequencies, potentially making the room sound thin. Carpets can make a room sound dull, although there are cases in which carpeting specific areas can be acoustically beneficial. The surface finishes used near a stage or the platform of a sanctuary are especially important acoustically.

In rooms where noise has been a problem, a renovation can provide an opportunity to correct faulty HVAC design, or to build in additional acoustical isolation into the walls. Often, noise problems are not recognized as such, but result instead as complaints about people being unable to understand speech in the rooms. An easy indication of an HVAC noise problem is to listen carefully from the back of the room while someone at the front of the room talks, first with the HVAC system off, then with it on. If speech intelligibility becomes an issue when the HVAC system is turned on, the HVAC noise is too great.

Correcting Poor Acoustics

There are many companies who sell acoustical treatment materials, as well as many people offering advice on the purchase of such materials. In some cases, this situation reminds one of the maxim "if your only tool is a hammer, every problem looks like a nail." While there are many acoustical problems that can benefit from acoustical absorption, an optimal solution always requires three elements: the right **amount** of the right **material** in the right **location**. Given a room that meets the statistical requirements, mathematical calculations can help in determining the right amount of material, but both other factors require specific expertise in acoustics to produce the best results.

However, we are getting ahead of ourselves. First, it is essential to determine in just what way the acoustics of the room in question are "poor". In some cases, no thought was given to acoustics when the room was originally designed. Usually, this leads to excessive RT60 and/or echo problems, although in more elaborate rooms, there can be focusing problems due to domed ceilings, curved rear walls, etc. Often such rooms exhibit excessive externally- or internally-generated noise. If there are echo problems, these can manifest as discrete ("slap") echoes, or as "flutter echoes" which are particularly disturbing to speech intelligibility and musical clarity. Echo problems can be remedied by using either absorptive materials (in the proper thickness), acoustical diffusion, or a combination of the two. If the RT60 is suitable, additional absorption is not needed, so diffusion is the answer.

Excessive RT60 is remedied by the addition of acoustical absorption of the proper thickness, applied in the appropriate locations. Sound-focusing problems are so varied in their exact causes and associated visual aesthetics that no general statements can be made about them. Noise problems should be addressed by an acoustical consultant experienced in HVAC design and noise isolation.

Other cases of "poor acoustics" can be expressed as "Music sounds thin" or "This room sounds harsh" or "Even when the sound system is loud enough, people in the back can't hear". Depending upon the room and the music style (acoustic or amplified), these comments can point to either sound system problems or acoustical problems. Some rooms have been built with improper acoustical treatment, resulting in lack of reverberation to support the music. Other rooms may have been quite good when first constructed, but incorrect maintenance procedures (such as painting acoustical panels) or the ravages of time (such as joints opening between boards in a ceiling) have hurt the acoustics.

Although it seems obvious, if the style of music performed in a room is changed, the optimum acoustics will change also. Thus a room that was known to have good acoustics can later be accused of having poor acoustics due to changes in program material.

Stage support for musicians and vocalists is another area that should be examined during planning for a renovation. Many venues are deficient in this area, and the remedies – properly shaped reflecting surfaces near the stage or platform – are much more easily accomplished as part of a renovation than as a "maintenance upgrade".

In any case, remediating poor acoustics is a job requiring specific skills and training, and may involve solutions that are radically different from those envisioned by non-specialists.

Appendix A: Obtaining Good Advice on Acoustics

There are many careers whose practitioners are affected by acoustics. In my career, I have seen cases in which musicians, music directors, organ builders, architects, music store salespeople, sound system technicians or salespeople, recording engineers, telephone technicians, and radio and TV announcers have been quoted as "experts" in acoustics. While different people have different levels of experience, I think we can agree that you would not go to a football coach or weight-lifter for medical advice, even though their work depends upon the working of the human body. When you need acoustical advice, your source should be someone who has the specialized education and experience needed to give you the correct advice. Advice that is wrong, or even "close to the mark", will not give you the hoped-for results, and may prove expensive.

How, then, do you decide whom to contact? First, be sure that you have defined your acoustical concern in your own mind. Then there are several ways to identify candidates to act as your consultant:

1. Ask a respected architect to recommend a good acoustical consultant.
2. Check the membership database of the National Council of Acoustical Consultants (NCAC), which, despite the name, is an international organization. Similar professional organizations exist in other countries, and they can be located on the Internet or by a librarian.
3. Ask a respected sound contractor for a recommendation.

By following these rules, you should be able to select several qualified candidates.

Next, you face the task of deciding which candidate to select. Again, there are a few rules:

1. Find out whether each consultant under consideration is a "real consultant". Real consultants work for the client, and do not have any financial interest in selling products. Many sound contracting companies sell what they call "consulting services". However, although in some cases these services may provide good advice, in many other cases the advice is skewed by desire to sell products or other services. The same can be said of "acoustical contractors", who usually sell and install acoustical treatment as well as non-acoustical wall and ceiling materials.

2. Check each candidate's educational background. Although this varies from country to country, in the US there are very few schools that offer degrees specifically in acoustics, so many consultants' degrees are in related fields such as Electrical Engineering, Mechanical Engineering, Physics, or Aeronautical and Aerospace Engineering. This unfortunate lack of specificity in degree names makes the next step even more important.

3. Secure a client list from those consultants whom you are considering, focusing especially upon venues similar in purpose and design to the one for which you need consulting services. Contact these clients to find out how similar their situations were to the one with which you are involved, and find out how well the consultant's advice succeeded in achieving the stated goals. Also find out how well the consultant worked with the owner's personnel and other stake-holders in the project.

4. Determine the number of years' experience each consultant has in the field of acoustics.

5. Find out what affiliations each candidate has with professional organizations such as the NCAC, Acoustical Society of America (ASA), Institute of Noise Control Engineering, and Audio Engineering Society. Although these are all international societies, there are also a number of societies for acoustics professionals with home offices in other nations besides the US. The reason that professional society membership is crucial is that formal education goes out-of-date quickly, and regular contact with other practicing professionals, plus exposure to research journals, is essential to keeping one's knowledge current. It is also helpful to know whether

the candidate has held office in professional societies, as this fact indicates something of the esteem in which the candidate is held by his/her peers.

A good consultant considers that (s)he is working for the ultimate client, which is usually the owner or tenant of the building whose acoustics are being considered. However, the financial arrangements through which the work occurs can take different forms. In many instances, the consultant contracts directly with the owner or tenant. In other cases, the contract is made through an architect or builder. Usually, the consultant's final report containing his/her advice for the project will be submitted to the person or firm through whom the financial arrangements are made. This is not always the best arrangement. For example, if a church contracts for acoustical consulting and sound system design, and then the church receives the consultant's report, care must be taken to ensure that the consultant's recommendations are incorporated into the architectural plans, or in the case of a design-built project not using architectural services, into the builder's plans. Failure to follow through on implementation of a consultant's recommendations will result in poor performance of the venue, and the money for the consultant's fee will have been wasted. When a consultant designs a sound system for a venue, usually the consultant collaborates with the owner, architect, and/or builder in selecting the contractor to whom is awarded the contract to install the system. Typically, the contractor is required to submit certain schedules and drawings to the consultant, and the system must pass the consultant's "commissioning" tests, as a provision of the installation contract. The consultant has no leverage to ensure that such contract provisions are enforced, since the installation contract is granted by the owner or builder. Therefore, it is customary for the consultant's approval of the complete installation to be obtained by the owner before the final payment is made to the installation company. Sometimes similar provisions are established for acoustical work based upon consultant recommendations.

When a consultant offers a proposal for a project, the proposal will cover the basic scope of work needed to address the client's expressed concerns. However, often a consultant will identify other possible areas of concern, and will include optional additions to the scope of the consultancy. These options should be carefully considered, since assuming that you have selected a good consultant, they are included based upon the consultant's experience – not based mainly on pecuniary considerations.

A complete consulting contract often includes a specific allowance for site visits and other meetings. It is important to remember that if this allowance is exceeded, the consultant may require an addendum to the contract, involving additional fees, to cover the added expenses. Clients who ask for too many meetings and site visits may be in for an unpleasant surprise, so it is important to use these functions only when and as needed.

Appendix B: Facts and Myths about Acoustical Analysis

As discussed early in this book, the quantitative analytical approach to architectural acoustics was fathered by Wallace Clement Sabine early in the 20th century. The methods pioneered by Sabine have been used by virtually all acousticians since his time, and have been found to provide predictions that agree well with the measured performance of acoustical spaces, provided that certain conditions are met. Basically, these conditions ensure that analysis based upon statistically likely behavior of sound (which is the basis of the Sabine approach) is valid in the room. This means that the room must have fairly equal distribution of acoustical absorption on all walls, floor, and ceiling, and that the room is a reasonably open space. Rooms having features such as deep underbalcony recesses or long narrow halls that act as coupled spaces do not meet the necessary conditions for the Sabine method to make valid predictions. Throughout the years since the Sabine theory was introduced, efforts have been made to modify the theory to account for rooms that do not meet these conditions. An example is the Fitzroy equation for predicting RT60 in a room in which most of the acoustical absorption is on one or two surfaces, such as an auditorium in which the audience provides most of the absorption, or a sparsely-occupied large meeting room with an absorptive ceiling. These efforts have met with mixed success.

In the late 20th century, sophisticated computer programs began to be introduced which analyzed rooms by "ray-tracing". This is an approach that assumes that the sound source sends out rays of sound, which, like rays of light, travel in straight lines, and which reflect off obstacles in a specular manner. This approach was much like the mechanism Sabine envisioned when he developed his system, but it has the advantage of accounting for uneven distribution of absorption. These ray-tracing programs have evolved into several pieces of commercially available acoustical analysis software sold under such names as EASE, CATT Acoustic, and Odeon. Virtually all acoustical consultants, and many sound-system consultants and contractors, use one or more of these programs.

The advantage of using a ray-tracing program is that it can provide several different ways of examining RT60, as well as giving numbers for clarity, definition, speech intelligibility, strength, stage support, and levels of direct and reflected sound. If used by persons with a good understanding of acoustic fundamentals, the results provided by such programs can lead to a useful recommendation for a venue much more quickly than would manual calculations or spreadsheet analysis. If used by persons without such understanding, the results can lead equally quickly to disastrous errors. Building the computer model of the room requires informed judgment as to what features of the architecture and furnishings are acoustically relevant. Modeling too much detail can create errors in results just as easily a can modeling with insufficient detail. Knowing what sort of quantitative results to expect is essential in order for the consultant to recognize "out-of-bounds" results that warrant further investigation as to the correctness of the model or the setup of the computer analysis.

Aside from the expertise of the user of a ray-tracing program, there is a fundamental limitation on the accuracy of such programs due to the assumptions built into the ray-tracing method itself. Neither sound waves nor light waves travel in straight lines under all conditions. Both kinds of waves require a more physically correct form of analysis – "wave analysis" – to provide accurate results. If a handclap (acoustically small source emitting mostly high frequencies) is reflected from a wall (acoustically large obstacle), the trajectory of the echo can be predicted pretty well by ray-tracing. But if low-frequency sound from a large loudspeaker encounters a 6" diameter vertical pillar, much of the sound will diffract around the pillar, leaving almost no shadow behind the pillar. In this case, ray-traced results will not be accurate. Even in a lecture hall, the important acoustics involve sounds having wavelengths from about 10' to 2", making diffraction important for many features of architecture and room furnishings. The first attempt to deal with ray-tracing's inability to account for wave effects such as diffraction was the addition of a "scattering coefficient" to the reflection algorithm used in ray-tracing analysis. This allowed the user to specify a certain percentage of rays that would be more-

or-less randomly scattered, rather than specularly reflected. The addition of frequency dependence to the scattering behavior was a further improvement. However, while scattering can fill in shadows that ray-tracing creates, but which diffraction would fill in an actual room, it cannot do this until the ray has reflected off a scattering surface. A sound wave will diffract upon first contact with an obstacle, not just after one or more reflections.

There are other wave phenomena in real rooms that are also not captured by ray-tracing. These include the complex behavior of sound waves as they travel across the heads and shoulders of an audience, the effects of temperature and humidity upon sound waves, resonances or "room modes", and interference effects resulting in frequency-response aberrations. Because of all these wave effects, ray tracing can provide useful, but not truly accurate results. In the hands of a properly educated, experienced user, the results provided by ray tracing programs can be very helpful, but for the inexperienced user, they can be deceptive.

One area of great advantage provided by ray-tracing analysis is mapping of sound level produced by a loudspeaker. This function offers far greater accuracy than could be achieved using the horizontal and vertical coverage angles specified on the data sheet of the loudspeaker. Loudspeaker manufacturers have agreed on two standard formats for presenting radiation of sound at various frequencies and angles: the **EASE** format and the **Common Loudspeaker Format** (**CLF**). An example of a sound map showing the sound level in the audience area from the horrid but unfortunately common "four corners" speaker layout is shown in Figure B-1.

Although this map shows that sound level is fairly even across the audience area, it also shows geometrical lines illustrating areas where constructive and destructive interference from the four speakers will make the sound irritating and unnatural. This sort of information is more difficult for both consultants and clients to comprehend if presented verbally rather than in the pictorial form provided by ray-tracing software.

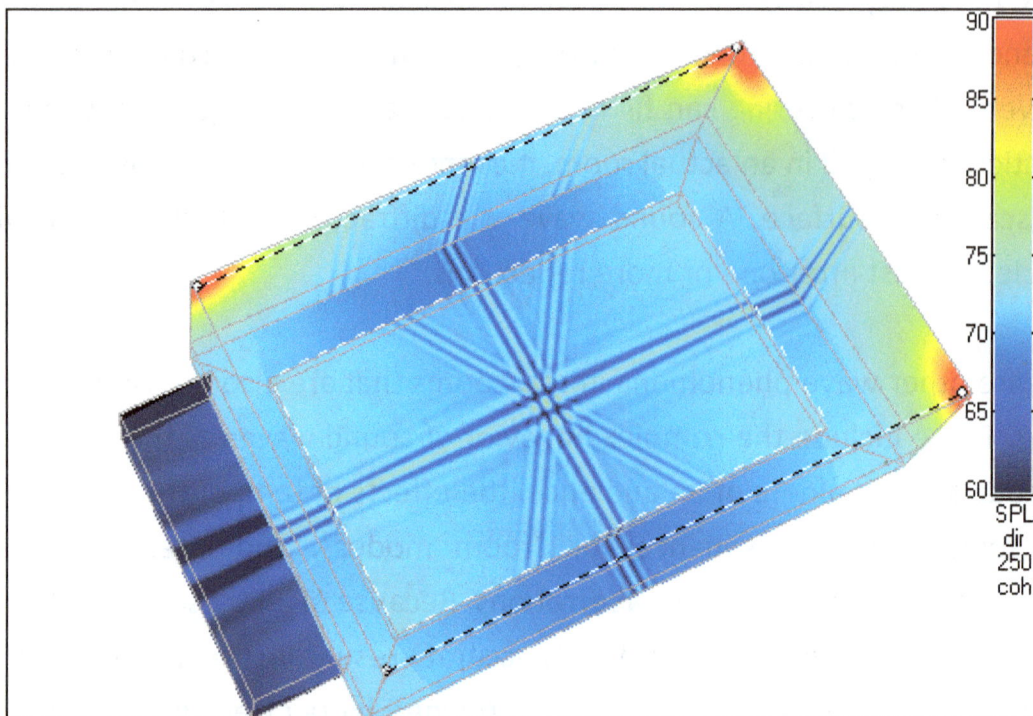

Figure B-1: Interference Produced by "4-Corners" Speaker Layout

One other useful function provided by ray-tracing software is "auralization". This is a process in which sound recorded in an anechoic chamber (RT60 close to zero) is digitally modified by a file resulting from the ray-tracing analysis of the room under consideration, to produce an audio experience that simulates the sound of the proposed room. The first attempts at auralization took immense amounts of computer time and produced mediocre results. Over the years, tweaks have been made to the modeling algorithms that now permit a much closer approximation of the actual room sound. Also, increases in computer speed by orders of magnitude have both made the auralization process faster and eliminated the need for certain shortcuts which were formerly used to reduce computer time.

One of the limitations on the usefulness of auralization is the available anechoic recordings for use as source files. It is difficult to find an anechoic chamber large enough for a symphony orchestra, and it is impossible to make a true anechoic recording of a pipe organ, so auralizations using either of these cannot be made. However, there a number of good anechoic recordings of male speech, female speech, and solo instruments including guitar, violin, cello, brasses, and woodwinds.

In the process of creating an auralization, the operator has the choice of ways in which it is to be played back. Although it is possible to create an auralization for playback through speakers, this approach is not very useful, both because of the variability of speakers – even of the same model – and because of the fact that the sound heard will include the acoustics of the listening room as well as the modeled room used for auralization. The better approach is to create the auralization for "binaural" playback. This is a method that incorporates a head-related transfer function to preserve the directional characteristics of the original sound, and involves the use of excellent headphones to exclude the effects of the listening room. In some cases, the auralization process allows for the behavior of specific brands and models of headphones, removing that variable from the listener's experience. Although different ray-tracing programs incorporate slightly different methods of auralization, and each company claims that their approach is superior, the fact is that there are small differences in the results, and that because of the fundamental limitations of ray-tracing, none of them can provide a rigorously accurate auralization.

In summary, then, as a user of acoustical spaces, what can you expect from computerized acoustical modeling? You can expect that any competent consultant will be proficient in the use of at least one acoustical modeling program, and that (s)he will use it to enhance the accuracy and efficiency of his/her work. You can expect that pictorial illustrations of the effects of proposed changes can in many cases be easily provided for your consideration. You can expect that an auralization of the proposed design can be made available for your hearing, and that the auralization will probably be closer to reality than will be your acoustical memory of the sound of any particular space. (Acoustical memory has been shown to be quite short: meaningful "A-B" comparisons should occur within at most a few minutes, or preferably, seconds.) But you cannot expect that ray tracing analysis will predict acoustical behavior with rigorous accuracy.

Appendix C: References

Chapter 2

Cavanaugh, William J., and Wilkes, Joseph A.: *Architectural Acoustics*, John Wiley & Sons, Inc., New York, 1999, p. 9.

Everest, F. Alton: *Master Handbook of Acoustics*, 4th Ed., McGraw Hill, New York, 2001, p. 32.

Knudsen, Vern O., *Architectural Acoustics*, John Wiley & Sons, Inc., New York, 1932, p. 419.

Chapter 3

Knudsen, Vern O., *Architectural Acoustics*, John Wiley & Sons, Inc., New York, 1932, p. 73.

Robinson, D. W.; and Dadson, R. S.: "A Re-determination of the Equal-Loudness Relations for Pure Tones", *British Journal of Applied Psychology*, **7** (1956), pp. 166-181.

Chapter 4

Fletcher, H.., "Physical Characteristics of Speech and Music," *Bell System Technical Journal*, **10**, p. 349 (July, 1931).

Knudsen, Vern O., *Architectural Acoustics*, John Wiley & Sons, Inc., New York, 1932, p. 103.

Shorter, D. E. L.; Manson, W. I.; and Wigan, E. R.: "The Subjective Effect of Limiting the Upper Audio Frequency Range", *E. B. U Review*, **57**, Part A (October, 1959).

Sivian, L. J.; Dunn, H. K.; and White, S. D.: "Absolute Amplitudes and Spectra of Certain Musical Instruments and Orchestras", *Bell System Technical Journal*, October, 1929.

Chapter 9

Honeycutt, Richard A., "Pews and Cushions and Carpets – Oh My!" *Sound and Vibration*, December, 2009, pp. 14-16.

Chapter 11

Honeycutt, Richard A., "Don't Let Your Multipurpose Room Sound Like a Train Station", *Church Executive*, August, 2010, pp. 34-35.

168